85 Springer Series in Solid-State Sciences
Edited by Klaus von Klitzing

Springer Series in Solid-State Sciences

Editors: M. Cardona   P. Fulde   K. von Klitzing   H.-J. Queisser

Managing Editor: H. K. V. Lotsch

Volumes 1–49 are listed on the back inside cover

50 **Multiple Diffraction of X-Rays in Crystals**
By Shih-Lin Chang

51 **Phonon Scattering in Condensed Matter**
Editors: W. Eisenmenger, K. Laßmann, and S. Döttinger

52 **Superconductivity in Magnetic and Exotic Materials**
Editors: T. Matsubara and A. Kotani

53 **Two-Dimensional Systems, Heterostructures, and Superlattices**
Editors: G. Bauer, F. Kuchar, and H. Heinrich

54 **Magnetic Excitations and Fluctuations**
Editors: S. Lovesey, U. Balucani, F. Borsa, and V. Tognetti

55 **The Theory of Magnetism II**
Thermodynamics and Statistical Mechanics
By D. C. Mattis

56 **Spin Fluctuations in Itinerant Electron Magnetism**   By T. Moriya

57 **Polycrystalline Semiconductors,**
Physical Properties and Applications
Editor: G. Harbeke

58 **The Recursion Method and Its Applications**
Editors: D. Pettifor and D. Weaire

59 **Dynamical Processes and Ordering on Solid Surfaces**
Editors: A. Yoshimori and M. Tsukada

60 **Excitonic Processes in Solids**
By M. Ueta, H. Kanzaki, K. Kobayashi, Y. Toyozawa, and E. Hanamura

61 **Localization, Interaction, and Transport Phenomena**
Editors: B. Kramer, G. Bergmann, and Y. Bruynseraede

62 **Theory of Heavy Fermions and Valence Fluctuations**
Editors: T. Kasuya and T. Saso

63 **Electronic Properties of Polymers and Related Compounds**
Editors: H. Kuzmany, M. Mehring, and S. Roth

64 **Symmetries in Physics**   Group Theory Applied to Physical Problems
By W. Ludwig and C. Falter

65 **Phonons: Theory and Experiments II**
Experiments and Interpretation of Experimental Results   By P. Brüesch

66 **Phonons: Theory and Experiments III**
Phenomena Related to Phonons
By P. Brüesch

67 **Two-Dimensional Systems: Physics and New Devices**
Editors: G. Bauer, F. Kuchar, and H. Heinrich

68 **Phonon Scattering in Condensed Matter V**
Editors: A. C. Anderson and J. P. Wolfe

69 **Nonlinearity in Condensed Matter**
Editors: A. R. Bishop, D. K. Campbell, P. Kumar, and S. E. Trullinger

70 **From Hamiltonians to Phase Diagrams**
The Electronic and Statistical-Mechanical Theory of sp-Bonded Metals and Alloys
By J. Hafner

71 **High Magnetic Fields in Semiconductor Physics**
Editor: G. Landwehr

72 **One-Dimensional Conductors**
By S. Kagoshima, H. Nagasawa, and T. Sambongi

73 **Quantum Solid-State Physics**
Editors: S. V. Vonsovsky and M. I. Katsnelson

74 **Quantum Monte Carlo Methods** in Equilibrium and Nonequilibrium Systems
Editor: M. Suzuki

75 **Electronic Structure and Optical Properties of Semiconductors**
By M. L. Cohen and J. R. Chelikowsky

76 **Electronic Properties of Conjugated Polymers**
Editors: H. Kuzmany, M. Mehring, and S. Roth

77 **Fermi Surface Effects**
Editors: J. Kondo and A. Yoshimori

78 **Group Theory and Its Applications in Physics**
By T. Inui, Y. Tanabe, and Y. Onodera

79 **Elementary Excitations in Quantum Fluids**
Editors: K. Ohbayashi and M. Watabe

80 **Monte Carlo Simulation in Statistical Physics**
An Introduction
By K. Binder and D. W. Heermann

81 **Core-Level Spectroscopy in Condensed Systems**
Editors: J. Kanamori and A. Kotani

82 **Introduction to Photoemission Spectroscopy**
By S. Hüfner

83 **Physics and Technology of Submicron Structures**
Editors: H. Heinrich, G. Bauer, and F. Kuchar

84 **Beyond the Crystalline State**
An Emerging Perspective
By G. Venkataraman, D. Sahoo, and V. Balakrishnan

85 **The Fractional Quantum Hall Effect**
Properties of an Incompressible Quantum Fluid
By T. Chakraborty and P. Pietiläinen

T. Chakraborty
P. Pietiläinen

The Fractional Quantum Hall Effect

Properties of an
Incompressible Quantum Fluid

With 85 Figures

Springer-Verlag Berlin Heidelberg New York
London Paris Tokyo

Dr. Tapash Chakraborty

Max-Planck-Institut für Festkörperforschung, Heisenbergstrasse 1
D-7000 Stuttgart 80, Fed. Rep. of Germany

Dr. Pekka Pietiläinen

Department of Theoretical Physics, University of Oulu,
Linnanmaa, 90570 Oulu 57, Finland

Series Editors:
Professor Dr., Dres. h. c. Manuel Cardona
Professor Dr., Dr. h. c. Peter Fulde
Professor Dr. Klaus von Klitzing
Professor Dr. Hans-Joachim Queisser

Max-Planck-Institut für Festkörperforschung, Heisenbergstrasse 1
D-7000 Stuttgart 80, Fed. Rep. of Germany

Managing Editor: Dr. Helmut K. V. Lotsch

Springer-Verlag, Tiergartenstrasse 17
D-6900 Heidelberg, Fed. Rep. of Germany

ISBN 3-540-19111-9 Springer-Verlag Berlin Heidelberg New York
ISBN 0-387-19111-9 Springer-Verlag New York Berlin Heidelberg

Library of Congress Cataloging-in-Publication Data. Chakraborty, Tapash, 1950- The fractional quantum Hall effect / Tapash Chakraborty, Pekka Pietiläinen. p. cm.–(Springer series in solid-state sciences ; 85) Bibliography: p. 1. Quantum Hall effect. I. Pietiläinen, Pekka, 1946-. II. Title. III. Series. QC612.H3046 1988 537.6'22–dc 19 88-24890

This work is subject to copyright. All rights are reserved, whether the whole or part of the material is concerned, specifically the rights of translation, reprinting, reuse of illustrations, recitation, broadcasting, reproduction on microfilms or in other ways, and storage in data banks. Duplication of this publication or parts thereof is only permitted under the provisions of the German Copyright Law of September 9, 1965, in its version of June 24, 1985, and a copyright fee must always be paid. Violations fall under the prosecution act of the German Copyright Law.

© Springer-Verlag Berlin Heidelberg 1988
Printed in Germany

The use of registered names, trademarks, etc. in this publication does not imply, even in the absence of a specific statement, that such names are exempt from the relevant protective laws and regulations and therefore free for general use.

Printing: Druckhaus Beltz, 6944 Hemsbach/Bergstr.
Binding: J. Schäffer GmbH & Co. KG., 6718 Grünstadt
2154/3150-543210 – Printed on acid-free paper

Foreword

QC
612
.H3
C46
1988

The experimental discovery of the fractional quantum Hall effect (FQHE) at the end of 1981 by Tsui, Störmer and Gossard was absolutely unexpected since, at this time, no theoretical work existed that could predict new structures in the magnetotransport coefficients under conditions representing the extreme quantum limit. It is more than thirty years since investigations of bulk semiconductors in very strong magnetic fields were begun. Under these conditions, only the lowest Landau level is occupied and the theory predicted a monotonic variation of the resistivity with increasing magnetic field, depending sensitively on the scattering mechanism. However, the experimental data could not be analyzed accurately since magnetic freeze-out effects and the transitions from a degenerate to a nondegenerate system complicated the interpretation of the data. For a two-dimensional electron gas, where the positive background charge is well separated from the two-dimensional system, magnetic freeze-out effects are barely visible and an analysis of the data in the extreme quantum limit seems to be easier. First measurements in this magnetic field region on silicon field-effect transistors were not successful because the disorder in these devices was so large that all electrons in the lowest Landau level were localized. Consequently, models of a spin glass and finally of a Wigner solid were developed and much effort was put into developing the technology for improving the quality of semiconductor materials and devices, especially in the field of two-dimensional electron systems.

The formation of a Wigner lattice has been observed for the two-dimensional electron gas at the helium surface with the consequence that all sorts of unexpected results on two-dimensional systems in semiconductors were assigned to some kind of charge-density-wave or Wigner crystallization. First attempts to explain the FQHE were therefore guided by the picture of a Wigner solid with triangular crystal symmetry. However, a critical analysis of the data demonstrated that the idea of the formation of an incompressible quantum fluid introduced by Laughlin seems to be the most likely explanation.

The theoretical work collected in this book demonstrates that the Laughlin wave function forms a very good basis for a discussion of the FQHE. Even though many questions in the field of FQHE remain unanswered, this book offers a valuable source of information and is the first general review of the work of different groups in this field. The intense activity in the field of high-T_c superconductivity also calls for a book about the FQHE since certain similarities seem to be emerging in the theoretical treatment of the quantum Hall effect and that of high-T_c superconductivity.

I hope that this book will inspire scientists to new ideas.

Stuttgart, June 1988   Klaus von Klitzing

Preface

In the field of the fractional quantum Hall effect, we have witnessed tremendous theoretical and experimental developments in recent years. Our intention here is to present a general survey of most of the theoretical work in this area. In doing so, we have also tried to provide the details of formal steps, which, in many cases, are avoided in the literature. Our effort is motivated by the hope that the present compilation of theoretical work will encourage a nonexpert to explore this fascinating field, and at the same time, that it will provide guidelines for further study in this field, in particular on many of the open problems highlighted in this review. Although the focus is on the theoretical investigations, to see these in their right perspective, a brief review of the experimental results on the excitation gap is also presented. This review is of course, by no means complete; the field continues to present new surprises, and more theoretical work is still emerging. However, we hope that the compilation in its present form will to some extent satisfy the need of the experts, nonexperts and the curious.

Stuttgart, Oulu,   Tapash Chakraborty
January, 1988   Pekka Pietiläinen

Acknowledgments

We wish to express our gratitude to Professor Peter Fulde and Professor Klaus von Klitzing for offering us the opportunity to write this review and for their encouragement and support during the course of this work. We are very much indebted to Professor von Klitzing for writing a foreword for this publication.

One of us (T. C.) would like to thank all his colleagues at the Max-Planck-Institute for their invaluable criticism and advice. Among others, he would like to thank particularly T. K. Lee, I. Peschel and G. Stollhoff for critically reading part of the manuscript and offering many suggestions for improvement, and I. V. Kukushkin, for providing some of his as yet unpublished experimental results. He also thanks B. I. Halperin, H. L. Störmer, A. H. MacDonald, S. M. Girvin, F. D. M. Haldane, F. C. Zhang and G. Fano for granting permission to reproduce some of the figures from their publications.

The other (P. P.) would like to thank Professor Fulde for arranging a visit to the Max-Planck-Institute, Stuttgart, during the final stage of the work. He also thanks the Department of Theoretical Physics, University of Oulu for support.

The manuscript was typeset by the authors using $\mathcal{A}_{\mathcal{M}}\mathcal{S}$-TeX. We would like to thank Gabriele Kruljac at the Computer Center of the Max-Planck-Institute, for her assistance in programming. We would also like to thank Dr. A. M. Lahee of Springer, whose careful reading of the manuscript and numerous suggestions have greatly improved the presentation.

Last but not the least, we would like to thank our families for their patience and understanding.

Contents

1. Introduction 1
2. Ground State 10
    2.1 Finite-Size Studies: Rectangular Geometry 10
    2.2 Laughlin's Theory 17
    2.3 Spherical Geometry 24
    2.4 Monte Carlo Results 29
    2.5 Reversed Spins in the Ground State 32
    2.6 Finite Thickness Correction 35
    2.7 Liquid-Solid Transition 37
3. Elementary Excitations 39
    3.1 Quasiholes and Quasiparticles 40
    3.2 Finite-Size Studies: Rectangular Geometry 49
    3.3 Spin-Reversed Quasiparticles 50
    3.4 Spherical Geometry 54
    3.5 Monte Carlo Results 56
    3.6 Experimental Investigations of the Energy Gap 65
    3.7 The Hierarchy: Higher Order Fractions 73
4. Collective Modes: Intra-Landau Level 83
    4.1 Finite-Size Studies: Spherical Geometry 83
    4.2 Rectangular Geometry: Translational Symmetry 85
    4.3 Spin Waves 94
    4.4 Single Mode Approximation: Magnetorotons 97
5. Collective Modes: Inter-Landau Level 109
    5.1 Filled Landau Level 109
    5.2 Fractional Filling: Single Mode Approximation 112
    5.3 Fractional Filling: Finite-Size Studies 117

6. Further Topics 121
   6.1   Effect of Impurities 121
   6.2   Higher Landau Levels 127
   6.3   Even Denominator Filling Fractions 131
   6.4   Half-Filled Landau Level in Multiple Layer Systems 137

7. Open Problems and New Directions 141

Appendices
   A   The Landau Wave Function in the Symmetric Gauge . . . 143
   B   The Hypernetted-Chain Primer 147
   C   Repetition of the Intra-Mode in the Inter-Mode 156

References 163

Subject Index 173

1. Introduction

The physics of an electron system in a magnetic field has gone through major developments in recent years following the spectacular discovery of the quantum Hall effect (QHE) by *von Klitzing* and his collaborators [1.1–3] and the subsequent discovery of the fractional quantum Hall effect (FQHE) [1.4–6]. The former effect has often been referred to as the integer QHE (IQHE). The major impetus in this field of research is due to the experimental realization of almost ideal two-dimensional (dynamically) electron systems. The electrons are dynamically two-dimensional because they are free to move in two spatial dimensions, and hence the wave vector is a good quantum number in two dimensions. In the third dimension, however, they have quantized energy levels. (In reality, the wave functions have a finite spatial extent in the third dimension [1.7]). The most widely used experimental systems are the two-dimensional electrons in the metal-oxide-semiconductor (MOS) inversion layer and the electron systems in GaAs-$Al_xGa_{1-x}As$ ($0 < x \leq 1$) heterostructures.

Conceptually, IQHE is quite simple. For a two-dimensional electron system in the x-y plane, subjected to a strong perpendicular magnetic field, application of an electric field along the x-direction induces a Hall current $j_y = -\sigma_{xy}E_x$ along the y-direction (Lorentz force). The classical Hall conductivity is then given as

$$\sigma_{xy} = -Nec/B \qquad (1.1)$$

with N being the electron concentration and c is the speed of light. At very low temperatures, however, Hall conductivity develops a series of plateaus with the conductivity given precisely by

$$\sigma_{xy} = -ie^2/h \qquad (1.2)$$

with i being an integer. This is a direct manifestation of Landau quantization of the two-dimensional electron system in a magnetic field. A qualitative understanding of IQHE is possible only in terms of noninteracting electrons (for a review of the theoretical and experimental results on the in-

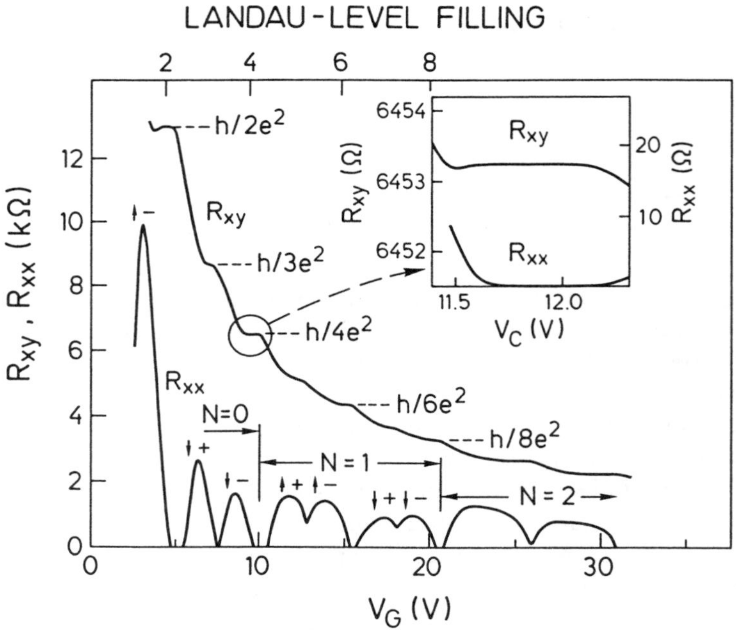

Fig. 1.1. The quantum Hall effect observed in Si(100) MOS inversion layer in a magnetic field of $B = 19$ T at temperature $T = 1.5$ K. The diagonal resistance and the Hall resistance are shown as a function of gate voltage ($\propto$ to electron concentration). The oscillations in R_{xx} are labelled by the Landau level index (N), spin ($\uparrow\downarrow$) and the valley ($\pm$). The upper scale indicates the Landau level filling described in the text, and the inset shows the details of a plateau for $B = 13.5$ T [1.2]

teger QHE, see [1.8]).[1] For an integer number i of completely filled Landau levels, the relation (1.2) is satisfied by the Hall resistance. The vanishing of diagonal resistance is due to the gaps in the single-particle density of states, and the presence of localized states in the energy gap is responsible for the width of the quantized Hall conductance plateaus.

In Fig. 1.1, we have reproduced the experimental results of *von Klitzing* [1.2] for a Si-MOSFET (metal-oxide-semiconductor field-effect transistor) inversion layer for a magnetic field of $B = 19$T. The diagonal resistance $R_{xx}(\approx \rho_{xx})$ and the Hall resistance $R_{xy}(\approx R_\mathrm{H})$ are plotted as a function of gate voltage V_g ($\propto$ electron concentration). The diagonal resistance is seen to vanish at different regions of V_g, indicating a current flow without any dissipation. In the same regions, R_{xy} develops plateaus and (1.2) is

---

[1] A brief review can also be found in [1.9]. Also, see the articles in [1.10].

obeyed with *extreme accuracy*. The experimental accuracy so far achieved is one part in 10^7. Another interesting feature of the above findings was that the quantization condition of the conductivity was very insensitive to the details of the sample (geometry, amount of disorder etc.). The quantized Hall resistance is more stable and more reproducible than any wire resistor and it is expected that from the year 1990, the quantized Hall resistance will be used as an international reference resistor [1.11].

Table 1.1. Observed fractions in the FQHE. The table was first presented by *Boebinger* in [1.17].

Filling factor	Accuracy of ρ_{xy} quantization	Structure in ρ_{xx}	References
$\frac{1}{3}$	3.0×10^{-5}	very strong minimum	[1.12]
$\frac{2}{3}$	3.0×10^{-5}	very strong minimum	[1.12]
$\frac{4}{3}$	9.0×10^{-4}	very strong minimum	[1.14,15]
$\frac{5}{3}$	1.0×10^{-3}	very strong minimum	[1.12,14,15]
$\frac{7}{3}$	–	weak minimum	[1.13,14]
$\frac{8}{3}$	–	weak minimum	[1.13–15]
$\frac{1}{5}$	–	weak structure	[1.12,13]
$\frac{2}{5}$	2.3×10^{-4}	very strong minimum	[1.12]
$\frac{3}{5}$	1.3×10^{-3}	very strong minimum	[1.12]
$\frac{4}{5}$	–	weak minimum	[1.12,16]
$\frac{7}{5}$	–	strong minimum	[1.16]
$\frac{8}{5}$	–	strong minimum	[1.16]
$\frac{2}{7}$	–	weak minimum	[1.4]
$\frac{3}{7}$	3.3×10^{-3}	strong minimum	[1.12]
$\frac{4}{7}$	–	strong minimum	[1.12,16]
$\frac{9}{7}$	–	strong minimum	[1.16]
$\frac{10}{7}$	–	strong minimum	[1.16]
$\frac{11}{7}$	–	strong minimum	[1.16]
$\frac{4}{9}$	–	weak structure	[1.12]
$\frac{5}{9}$	–	weak structure	[1.12,16]
$\frac{13}{9}$	–	weak minimum	[1.16]

The FQHE was discovered in a two-dimensional electron system developed in high-mobility GaAs-(AlGa)As heterostructures with very low amounts of disorder, at low temperatures, and in very high magnetic fields. Under these extreme conditions, the Hall resistance was found to be quantized to $\rho_{xy} = h/ie^2$ where i is now a simple rational *fraction*.[2] In the initial experiment [1.4], Hall resistance quantization was observed at $\frac{1}{3}$ and $\frac{2}{3}$. In subsequent studies [1.5,12] several other fractions were observed, where quantization occurs with very high accuracy. Table 1.1 presents a list of the fractions observed so far. The diagonal resistivity was also found to approach zero (Fig. 1.2) at the magnetic fields where the plateaus were observed, indicating, as in the case of IQHE, the existence of a gap in the

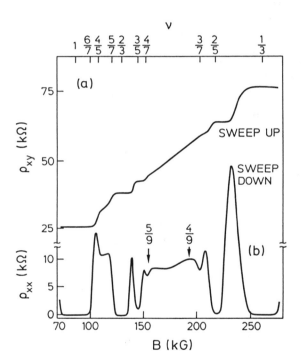

Fig. 1.2. The FQHE as observed in a GaAs-heterojunction. The observed fractions are $\frac{1}{3}, \frac{2}{3}, \frac{2}{5}, \frac{3}{5}, \frac{3}{7}$ and $\frac{4}{7}$. Weak structures were found to appear in ρ_{xx} near $\frac{4}{9}$ and $\frac{5}{9}$. The upper scale indicates Landau level filling defined in the text [1.12]

---

[2] The resistivity component ρ_{xx} is directly proportional to the conductivity component σ_{xx}, $(\sigma_{xx} = \sigma_{yy}, \sigma_{xy} = -\sigma_{yx})$, $\sigma_{xx} = \rho_{xx}/(\rho_{xx}^2 + \rho_{xy}^2)$. Therefore, $\rho_{xx} = 0$ in a plateau implies that $\sigma_{xx} = 0$ simultaneously, as long as $\rho_{xy} \neq 0$. Furthermore, the relation $\sigma_{xy} = -\rho_{xy}/(\rho_{xx}^2 + \rho_{xy}^2)$ implies that, in a plateau where ρ_{xx} vanishes, $\sigma_{xy} = -1/\rho_{xy}$ is also a constant.

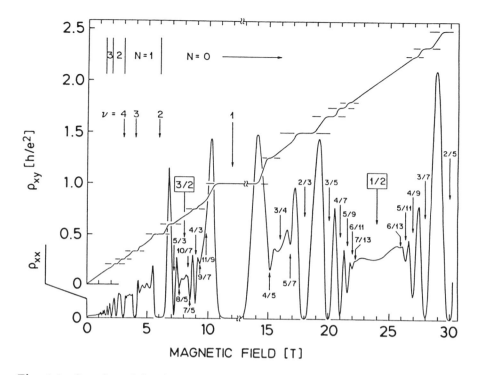

Fig. 1.3. Overview of the observed fractions in the FQHE measurements. The Landau level filling factor ν has been defined in the text [1.18]

excitation spectrum. The striking feature of the above results is that all fractional fillings appear with *odd* denominators. In Fig. 1.3, we have reproduced the latest results on the FQHE, where several new (as well as established) fractions have been resolved. FQHE has also been observed in high-mobility Si-MOSFET samples.[3]

In explaining the FQHE, the system of noninteracting electrons is, however, inadequate. According to our present understanding of FQHE, electron correlations play a major role in this effect, and there have been a variety of theoretical attempts to understand this unique many-electron phenomenon. In this volume, we have attempted to survey most of these theoretical approaches, and have tried to present in detail the current state of our understanding of this fascinating effect.

There has been another surprise recently in FQHE. In the higher Landau level (n=1), *Willet* et al. [1.18] have discovered a fractional Hall plateau at

---

[3] We have not attempted to review the experimental work on FQHE by various groups; these can be found e.g. in [1.6] and [1.17].

$\rho_{xy} = \left(\frac{h}{e^2}\right)/\frac{5}{2}$, corresponding to an *even* denominator filling $(= 2+\frac{1}{2})$. This particular finding will be discussed in Sect. 6.3.

As the fractional Hall steps are observable only in samples of very high mobility, impurity potentials are not expected to be very important in comparison with the electron-electron interactions. The first step in the explanation of FQHE would therefore be to study the properties of a system of two-dimensional interacting electrons in a uniform positive background with the magnetic field strength such that only the lowest Landau level is partially filled. One could then consider the effect of impurities as a perturbation.

In Chap. 2, we survey various theoretical methods to study the ground state properties of a system such as that described above. In all the calculations, the dimensionless density of the electrons is expressed as the *filling factor* of the Landau level,

$$\nu \equiv N_e \Phi_0/B = 2\pi \ell_0^2 \rho \qquad (1.3)$$

where N_e is the number of electrons, $\Phi_0 = \frac{hc}{e}$ is the magnetic flux quantum, the magnetic length ℓ_0 is defined as

$$\ell_0 \equiv \left(\frac{\hbar c}{eB}\right)^{\frac{1}{2}}, \qquad (1.4)$$

and ρ is the number density of electrons in the system. The unit of potential energy is $\frac{e^2}{\epsilon \ell_0}$, which is taken to be the energy scale throughout. Here, ϵ is the background dielectric constant. For magnetic fields of $B \gtrsim 10T$, where FQHE is generally observed, using the values $\epsilon \approx 12.9$ and $m^* \approx 0.067 m_e$ which are appropriate for GaAs, it is easy to verify that $\frac{e^2}{\epsilon \ell_0} \lesssim \hbar \omega_c$, the latter being the cyclotron energy (defined in Sect. 2.1). The admixture of states in higher Landau levels can thus be safely ignored as a first approximation.

In the extreme quantum limit (no Landau level mixing), the Hamiltonian describing the two-dimensional electron gas can be written as

$$\mathcal{H} = \frac{1}{2} \sum_q V(q) \left[\bar{\rho}(q)\bar{\rho}(-q) - \rho e^{-q^2 \ell_0^2}\right] \qquad (1.5)$$

where $\bar{\rho}(q)$ is the projected density operator (discussed in Sect. 4.4) and ρ is the average density of the particles in the system. In the absence of kinetic energy, the ground state is expected to be a solid — which means that the Hamiltonian describes a set of interacting classical particles. The operators $\bar{\rho}(q)$ do not commute with one another. However, in the limit $B \to \infty$ [1.19] we get

$$[\bar{\rho}_q, \bar{\rho}_{-q}] \to \ell_0 \qquad (1.6)$$

and one obtains a solid which is presumably triangular in the limit of infinite magnetic field. The earlier attempts to explain FQHE were mostly centered on the crystal state calculations. Various crystal state calculations [1.20,21] however, did not find any singularity at $\nu = \frac{1}{3}$. It is also difficult conceptually to understand how the crystal could carry electric current with no resistive loss, since the charge-density wave (CDW) would be pinned by the impurities. Furthermore, if the CDW is not allowed to move with a drift velocity $\boldsymbol{E} \times \boldsymbol{B}c/B^2$, the electron contribution to the Hall conductivity would vanish [1.10,22]. While there were some claims that a crystal-type trial wave function has a significantly lower energy than that of a liquid state [1.23–25], we beleive that a majority of the theoretical work to be described in this review has clearly established that the ground state of the electron system is a translationally invariant *liquid* state. The major step in arriving at this conclusion was made by *Laughlin* [1.26]. Various other calculations strongly support this conclusion.

In our attempt to explain the fractional Hall steps, we have to understand the exceptionally stable states of the electron system at particular rational values of ν. The pinning of the density at those values of ν would require that the energy versus density curve shows a cusp-type behavior. A cusp would imply a discontinuity of chemical potential, which would in turn, mean that the electron system is, in fact, *incompressible*[4] at those stable states. *Laughlin* introduced a quite radical concept at this point [1.26], proposing that the lowest energy charge excitations in the system are *fractionally* charged quasiparticles and quasiholes. Slight deviations from a stable ν would create those quasiparticles and quasiholes, costing a finite amount of energy. In Chap. 3, we discuss in detail, how these charged excitations are created and the various methods to compute their creation energies. We also review briefly some of the interesting experimental work on the energy gap from thermal activation of the diagonal resistivity, as well as from spectroscopic methods.

Once the lowest energy charged excitations are identified, the next natural step would be to study the lowest lying neutral excitations—the *quasiexcitons*. Finite-size studies have shown that a collective mode exists in the incompressible state with a finite gap, and a minimum at a finite wave vector. The mode is also well separated from the continuum. The minimum in the collective excitation spectrum is, in fact, similar to the *roton* minimum in liquid ^{4}He. These topics will be discussed in Chap. 4.

---

[4] This means that changing the area at constant magnetic field and electron number requires an an energy of $\delta E = E_g/[2\pi \ell_0^2 m]|\delta A|$, where m is an odd integer and E_g is the quasiparticle–quasihole energy gap to be described in Chap. 3, while for a compressible elastic medium, $\delta E \propto (\delta A)^2$.

When the Landau levels are completely filled or completely empty, the excitations in the absence of interactions, involve promoting an electron from an occupied state in the nth Landau level to an unoccupied state in the n'th landau level. The corresponding energy is simply the kinetic energy difference between the two levels. In the presence of interactions, one can obtain an expression for the dispersion which is exact to the lowest order in $\frac{e^2}{\epsilon \ell_0}/\hbar\omega_c$. One can also obtain the spin wave spectrum exactly. The dispersion can be studied indirectly through the cyclotron-resonance line shape in the weak-disorder limit. These results are discussed briefly in Chap. 5. In the case where the Landau levels are partially filled, the intra-Landau level mode influences the inter-Landau level mode. This interesting problem has been studied recently by some authors for the infinite systems as well as for finite-size systems, and also discussed in this chapter.

Finally, in Chap. 6, we survey various interesting effects, which are very important for our understanding of the FQHE in general, and how they influence the incompressible fluid state. There have been some studies of the effect of impurities on the incompressible fluid state using a Laughlin-type approach, as well as the finite-size systems. These investigations are far from complete, however. Another interesting problem is the FQHE at the higher Landau levels which has been given attention by various authors and is discussed in Chap. 6. Finally, in this chapter, we also discuss the recent developments concerning the FQHE for even denominator filling fractions. The theoretical work on these filling fractions is not so exhaustive compared to the vast amount of effort devoted to the odd denominator fillings. The experimental and theoretical results are just emerging. In Sects. 6.3 and 6.4, we have tried to present a brief picture of the present state of our understanding of this phenomenon.

In Appendix A, we present briefly the algebra for the single electron eigenstates in the symmetric gauge, relevant for the Laughlin type of approach. Appendix B contains a brief background on the hypernetted-chain theory, which is essential for understanding the quantitative results for the ground state and elementary excitations in Laughlin's approach. In Appendix C, we present some mathematical details on the magnetoplasmon modes discussed in Chap. 5.

Besides Laughlin's theory for the fractional quantization, there have also been other attempts to describe the effect. *Tao* and *Thouless* [1.27] proposed a method in the Landau gauge in which it was suggested that every third level in the degenerate Landau levels would be occupied in the unperturbed ground state, while the two intervening levels are empty. This could be viewed as a one-dimensional solid in the space of Landau levels. Later, *Thouless* [1.28] noticed that the wave function in their scheme includes long-range correlations. Thouless also found that such a long-range order is absent in Laughlin's wave function. He then argued that, since the odd denominators for fractional quantization come out naturally from Laughlin's theory and since the overlap of the Laughlin wave function with the exact ground state from finite-size calculations are remarkably large, whereas not much could be said in favor of the Tao-Thouless theory, the latter theory should be abandonned. We will not discuss this theory any further in the present review.

Another theoretical approach which will not be elaborated on is the *cooperative ring exchange* mechanism of *Kivelson* et al. [1.29,30]. Using a path integral formalism, these authors found that the important processes are those which involve cyclic exchange of large rings of electrons. While each individual ring makes only a small contribution to the energy, the number of available rings increases exponentially with the length of the ring. For fractional filling of Landau levels with odd denominators, the exchange energies of large rings add coherently and produce a downward cusp in the ground state energy; for other values of the filling fraction, the rings will contribute incoherently. Recently, *Thouless* and *Li* [1.31] studied this mechanism in a simple model where each ring is confined to a narrow channel. They found that for fractional filling of the Landau level, the exchange energy tends to give an *upward* cusp in the energy, which would imply that the fundamental assumption of the explanation of FQHE in terms of ring exchange is highly implausible. *Kivelson* et al. have recently argued [1.32] that the high-magnetic field is essential to obtain the correct results. This theory however needs to be developed further in order to be able to quantitatively compare its predictions with the available results from other theoretical work. In any case, as *Haldane* pointed out [1.33], this alternative approach, in order to be correct, has to be equivalent to Laughlin's theory.

2. Ground State

As mentioned previously, in the earlier investigations of FQHE, the crucial question was the nature of the ground state. As the GaAs-heterostructures where the FQHE was discovered have very high electron mobility, and because of the subsequent discovery of FQHE only in high-mobility Si-MOSFETs, the Coulomb interaction between electrons was quite naturally expected to play a dominant role in the FQHE. The earliest numerical calculation of the ground state including Coulomb interactions for various filling fractions was by *Yoshioka* et al. [2.1]. They investigated the eigenstates of an electron system in a periodic rectangular geometry, by numerically diagonalizing the Hamiltonian. Their results, as we shall discuss in the following section, revealed several interesting features; the most important result was, of course, that the ground state had a significantly lower energy than that of a Hartree-Fock (HF) Wigner crystal.

The major breakthrough in this problem was made by *Laughlin* [2.2–4], who proposed a *Jastrow-type* trial wave function for $\nu = \frac{1}{m}$ filling factor with m an *odd* integer. Based on this wave function, he also proposed the low-lying elementary excitations to be quasiparticles and quasiholes of fractional charge. Laughlin's work, as we shall try to demonstrate below, has an enormous influence in this field, and has been a major source of intuition for most of the theoretical studies of FQHE that followed.

In the following sections, we describe the approach of Yoshioka et al., and that of Laughlin in detail, and compare the results with very accurate Monte Carlo evaluation of the ground state properties obtained by *Levesque* et al. [2.5] and by *Morf* and *Halperin* [2.6,7]. Finite-size system results in the spherical geometry [2.8,9] are also presented for comparison.

2.1 Finite-Size Studies: Rectangular Geometry

Let us begin with the problem of a charged particle in a uniform magnetic field. Taking the z-direction to be along the field, the gauge is chosen such that the vector potential $\boldsymbol{A}$ has only one nonvanishing component, $A_y = Bx$

(or equivalently, $A_x = -By$) (Landau gauge). The other choice, viz. the symmetric gauge, will be dealt in Appendix A. For the motions in the x-y plane, the Hamiltonian is then simply,

$$\mathcal{H} = (\Pi_x{}^2 + \Pi_y{}^2)/2m \\ = p_x{}^2/2m + (p_y - eBx/c)^2/2m \tag{2.1}$$

where $\Pi = p - eA/c$ is the kinetic momentum. The variables are easily separable, and an eigenfunction is written in the form

$$\phi = e^{ik_y y}\chi(x). \tag{2.2}$$

The time-independent Schrödinger equation is then written as

$$\frac{\hbar^2}{2m}\left[-\frac{\partial^2}{\partial x^2} + \left(k_y - \frac{x}{\ell_0^2}\right)^2\right]\chi(x) = E\chi(x) \tag{2.3}$$

where the usual identification is made, $p_y = -i\hbar\partial_y \to \hbar k_y$. Equation (2.3) can be reexpressed in a more familiar form

$$\left[\frac{p_x^2}{2m} + \frac{1}{2}m\omega_c^2(x - x_0)^2\right]\chi(x) = E\chi(x)$$

where, $x_0 = k_y \ell_0^2$ with ℓ_0 defined in (1.4). The above equation is easily recognized as the Schrödinger equation corresponding to a *harmonic oscillator* of spring constant $\hbar^2/m\ell_0^2$, with equilibrium point at x_0. The natural frequency of such an oscillator is

$$\omega_c = \frac{eB}{mc} \tag{2.4}$$

which is the cyclotron frequency of a classical particle in the field.

The energy eigenvalues (the *Landau levels*) are then obtained from the quantum mechanical results for a harmonic oscillator:

$$E_n = \left(n + \frac{1}{2}\right)\hbar\omega_c. \tag{2.5}$$

The degeneracy of each Landau level is infinite since the energy is independent of k_y, and the functions corresponding to different values of k_y are linearly independent. When the system is confined in a rectangular cell with sides L_x and L_y, the degeneracy is in fact the number of allowed values of k_y, such that the center x_0 lies between 0 and L_x. With use of periodic boundary conditions we get, $k_y = 2\pi n_y/L_y$, with n_y an integer. The allowed values of n_y are then determined by the condition,

$$x_0 = \frac{2\pi n_y}{L_y}\ell_0^2, \quad 0 < x_0 < L_x. \tag{2.6}$$

The degeneracy N_s can then be expressed in terms of the magnetic length ℓ_0 as

$$N_s = \frac{L_x L_y}{2\pi \ell_0^2}. \tag{2.7}$$

Equation (2.7) can also be reexpressed in terms of the magnetic flux Φ and the flux quantum Φ_0 as

$$N_s = \frac{e}{hc}\Phi = \frac{\Phi}{\Phi_0}. \tag{2.8}$$

The Landau level degeneracy is thus the total number of flux quanta in the external magnetic field.

The eigenfunctions (ignoring the normalization factor) are now written

$$\phi = e^{ik_y y} \exp\left[-(x-x_0)^2/2\ell_0^2\right] H_n\left[(x-x_0)/\ell_0\right] \tag{2.9}$$

with H_n the Hermite polynomial. The functions are extended in y and localized in x. The localization remains unaffected under a gauge transformation.

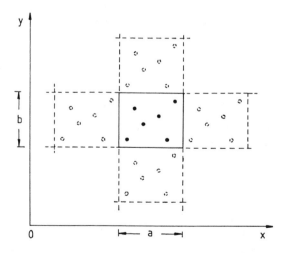

Fig. 2.1. Geometry considered by *Yoshioka* et al. [2.1]. Electrons (drawn as solid circles) are in a rectangular cell in the x-y plane. Because of the periodic boundary conditions, the electrons not only interact among themselves, but also with their images (drawn as open circles)

For the lowest Landau level, writing the center coordinate of the cyclotron motion as

$$X \equiv x_0 = k_y \ell_0^2 \tag{2.10}$$

the single-particle wave functions are (except for the normalization factor),

$$\phi = \exp\left[iXy/\ell_0^2 - (X-x)^2/2\ell_0^2\right]. \tag{2.11}$$

Following *Yoshioka* et al. [2.1], let us now put a finite number of electrons in the cell and introduce interactions among them.

Consider the situation shown in Fig. 2.1, where there are a few electrons in a rectangular cell of sides a and b, and a strong magnetic field perpendicular to the x-y plane. The electrons are considered to be in the lowest Landau level, and are spin polarized. Applying periodic boundary conditions in the y-direction, one obtains, $k_y = X_j/\ell_0^2 = 2\pi j/b$ for an integer j. For the periodic boundary condition along the x-direction, let us write, for an integer m, $X_m = a$. Clearly, $ab = 2\pi \ell_0^2 m$, and from (2.7), m is the Landau level degeneracy. Therefore, we have m different single-electron states in the cell with wave functions given by

$$\phi_j(\mathbf{r}) = \left(\frac{1}{b\sqrt{\pi}\ell_0}\right)^{\frac{1}{2}} \sum_{k=-\infty}^{+\infty} \exp\left[i\left(X_j + ka\right)y/\ell_0^2 - (X_j + ka - x)^2/2\ell_0^2\right] \tag{2.12}$$

with $1 \leq j \leq m$. The Coulomb interaction is written in the form

$$V(\mathbf{r}) = \frac{1}{ab}\sum_q \frac{2\pi e^2}{\epsilon q}\exp(i\mathbf{q}\cdot\mathbf{r}) \tag{2.13}$$

where $\mathbf{q} = \left(\frac{2\pi s}{a}, \frac{2\pi t}{b}\right)$ and s and t are integers. The Hamiltonian is now,

$$\mathcal{H} = \sum_j W a_j^\dagger a_j + \sum_{j_1}\sum_{j_2}\sum_{j_3}\sum_{j_4} A_{j_1 j_2 j_3 j_4} a_{j_1}^\dagger a_{j_2}^\dagger a_{j_3} a_{j_4} \tag{2.14}$$

where $a_j(a_j^\dagger)$ is the annihilation (creation) operator for the jth state. The single-electron part (interaction between an electron and its image) is a known constant [2.1],

$$W = -\frac{e^2}{\epsilon\sqrt{ab}}\left[2 - {\sum_{l_1 l_2}}' \varphi_{-\frac{1}{2}}\left\{\pi\left(\lambda l_1^2 + \lambda^{-1} l_2^2\right)\right\}\right] \tag{2.15}$$

where λ is the aspect ratio, l_1, l_2 are integers. The prime on the summation indicates that the term with $l_1 = l_2 = 0$ is excluded, and

$$\varphi_n(z) \equiv \int_1^\infty dt\, e^{-zt}\, t^n.$$

The two-electron part is given by

$$\begin{aligned}A_{j_1 j_2 j_3 j_4} &= \frac{1}{2}\int dr_1 dr_2 \phi^*_{j_1}(r_1)\phi^*_{j_2}(r_2)V(r_1-r_2)\phi_{j_3}(r_2)\phi_{j_4}(r_1) \\ &= \frac{1}{2ab}{\sum_q}' \sum_s \sum_t \delta_{q_x,2\pi s/a}\delta_{q_y,2\pi t/b}\delta'_{j_1-j_4,t}\frac{2\pi e^2}{\epsilon q} \\ &\quad \times \exp\left[-\frac{1}{2}\ell_0^2 q^2 - 2\pi is(j_1-j_3)/m\right]\delta'_{j_1+j_2,j_3+j_4}.\end{aligned} \quad (2.16)$$

Here the Kronecker delta with prime means that the equation is defined modulo N_s (or m), and the summation over q excludes $q_x = q_y = 0$.

For N_e electrons in the cell, the filling factor is $\nu = N_e/N_s$. The basis is specified by the occupation of the single-electron state: $(j_1, j_2, \ldots, j_{N_e})$. The total number of bases is $\binom{N_s}{N_e}$. In the absence of Coulomb interaction, all these bases are degenerate. The Coulomb interaction lifts the degeneracy and mixes the bases. Therefore, one needs to diagonalize rather large Hamiltonian matrices.

In order to simplify the calculation, one exploits the symmetry of the system. The total momentum in the y-direction, $J \equiv \sum_{i=1}^{N_e} j_i \pmod{N_s}$, is conserved due to the translational symmetry along the y-axis. For each J, the matrix dimension is $\sim \frac{1}{N_s}\binom{N_s}{N_e}$. Two values of J which differ by a multiple of N_e are equivalent due to translational symmetry along the x-axis. Therefore when N_s and N_e have no common factor, every eigenvalue is at least N_s-fold degenerate. However, if N_s and N_e do have a common factor, the states are less degenerate and the ground state is realized only for certain values of J. As an example, for $N_e = 4$ and $N_s = 12$, the three-fold ground state appears for $J = 2, 6$, and 10. Finally, due to the rotational symmetry, the cell with aspect ratio a/b is equivalent to that with aspect ratio b/a.

In Fig. 2.2, we present the numerical results of Yoshioka et al. [2.1], for the ground state energies per particle as a function of filling fraction in the lowest Landau level. There are several interesting features noticeable in the result. Let us first consider the case $\nu = \frac{1}{3}$. The ground state energy per particle for four-, five- and six-electron systems are (in units of $e^2/\epsilon \ell_0$): $-0.4152, -0.4127$, and -0.4129 respectively, and are extremely insensitive

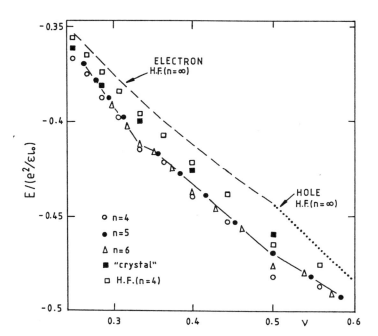

Fig. 2.2. Energies per particle for finite electron systems in a periodic rectangular geometry, as a function of filling factors in the lowest Landau level. The dashed and dotted lines are the energy of the electron and hole crystals within the HF approximation for the infinite system. Open circles, closed circles, and triangles are the results for N_e=4, 5 and 6 electrons for $\nu \leq \frac{1}{2}$ and holes for $\nu > \frac{1}{2}$. Closed squares denote the crystal state energies for the N_e=4 system. Open squares show the energy of the crystal state for the N_e=4 system obtained in the HF approximation. The solid line drawn through the N_e=5 ground state energies is a guide to the eye only [2.1]

to the system size. As we shall see in the next section, these results are also very close to the infinite system result.

The ground state energies tend to have downward *cusps* for $\nu = \frac{1}{3}$ and $\frac{2}{5}$. The system size is however too small to realize them more clearly. As discussed in Chap. 3, a cusp at the observed filling fractions is quite naturally required, in order to describe the incompressible fluid state proposed by Laughlin.

In Fig. 2.2, Yoshioka et al. also presented the energy of the crystal state for $N_e = 4$ system obtained in the exact diagonalization (closed squares) and HF approximation (open squares). The dashed and dotted lines show the energy of the electron and hole crystals resulting from the HF approximation for the infinite system [2.10]. These crystal state results show a smooth behavior at $\nu = \frac{1}{3}$ and are not the lowest energy state.

In order to investigate the eigenstates of the Hamiltonian, *Yoshioka* et al. [2.11,12] calculated the pair distribution function,

$$g(r) \equiv \frac{ab}{N_e(N_e-1)} \langle \Psi | \sum_{i \neq j} \delta(r + r_i - r_j) | \Psi \rangle$$

$$= \frac{1}{N_e(N_e-1)} \sum_q \sum_{j_1} \cdots \sum_{j_4} \exp\left[iq \cdot r - \frac{1}{2}q^2 \ell_0^2 - i(j_1 - j_3)\frac{q_x a}{m}\right]$$

$$\times \delta'_{j_1-j_4, q_y b/2\pi} \langle \psi | a_{j_1}^\dagger a_{j_2}^\dagger a_{j_3} a_{j_4} | \psi \rangle \tag{2.17}$$

where $|\Psi\rangle$ is one of the eigenstates. In Fig. 2.3, we reproduce the $g(r)$ at $\nu = \frac{1}{3}$ for the four-electron system studied by *Yoshioka* et al. [2.11]. The results are for the ground state and one of the excited states. For the excited state, $g(r)$ is almost identical to that of the triangular CDW state obtained from the HF approximation for the infinite system. Therefore these authors identified the state with the CDW state. On the other hand, the ground state $g(r)$ is found to be quite different. It has peaks at $r = (\pm\frac{a}{2}, 0)$ and $(0, \pm\frac{b}{2})$, but *not* at $r = (\pm\frac{a}{2}, \pm\frac{b}{2})$, where peaks would be expected if the state were a square CDW state. Furthermore, the shape of the peaks is different from the gaussian peak of the CDW state. It was therefore concluded that the ground state is a liquid-like state. The ground state $g(r)$ for higher values of N_e also shows the same structure [2.12].

In the lowest Landau level, there is electron-hole symmetry. The system with N_e electrons in N_s sites is thus equivalent to that with $(N_s - N_e)$ holes in N_s sites. When we choose the products of the single-electron eigenstates as a basis, the off-diagonal matrix elements for $\nu = N_e/N_s$ are same as those for $\nu = (1 - N_e/N_s)$ filling fractions for the same J values. The diagonal matrix elements differ only by a constant ΔE given by

$$\Delta E = \sum_{j_1=1}^{N_e} \sum_{j_2=1}^{N_e} (\mathcal{A}_{j_1 j_2 j_2 j_1} - \mathcal{A}_{j_1 j_2 j_1 j_2}) - \sum_{j_1=1}^{N_s-N_e} \sum_{j_2=1}^{N_s-N_e} (\mathcal{A}_{j_1 j_2 j_2 j_1} - \mathcal{A}_{j_1 j_2 j_1 j_2}).$$
(2.18)

The results for $\nu > \frac{1}{2}$, shown in Fig. 2.2 are obtained by the above arguments.

While these numerical results demonstrate that the ground state is not crystalline, as *Halperin* pointed out [2.13], not much insight is gained about the ground state wave function. The crucial step in obtaining a ground state wave function for a translationally invariant liquid state, and the mechanism for stabilizing the system at particular densities, was made by Laughlin, and is described below.

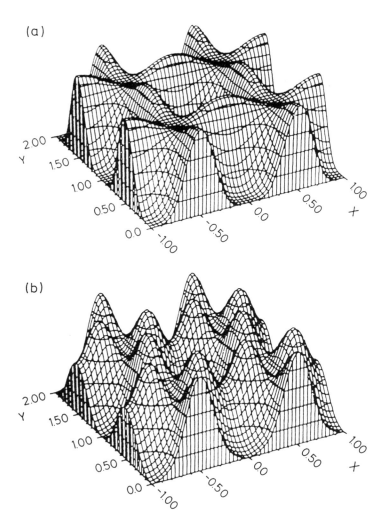

Fig. 2.3. Perspective view of the pair correlation function $g(r)$ obtained in [2.11] for $N_e = 4$ and $N_s = 12$; (a) the ground state (b) an excited state identified as the CDW state. The axes are normalized by the dimension of the cell: $X = x/a$ and $Y = y/b$

2.2 Laughlin's Theory

The original paper by *Laughlin* [2.2] contains the seminal ideas which have been elaborated since by him [2.3], and by other authors [2.6,13]. The reader is also referred to a recent article by *Laughlin* [2.4] for an insight on how the theory was orginally conceived. In the following, we discuss briefly a few key points of this theory in order to connect them with the other work discussed in this review.

Electrons are considered as usual, to be confined in the x-y plane and subjected to a magnetic field perpendicular to the plane. Considering the symmetric gauge vector potential, $\boldsymbol{A} = \frac{1}{2}B(y\hat{\boldsymbol{x}} - x\hat{\boldsymbol{y}})$, it is convenient to regard the x-y plane as a complex plane. For the lowest Landau level, the single-particle wave functions are eigenfunctions of orbital angular momentum (see Appendix A),

$$\phi_m(z) \equiv |m\rangle = \frac{1}{(2\pi \ell_0^2 2^m m!)^{\frac{1}{2}}} \left(\frac{z}{\ell_0}\right)^m e^{-|z|^2/4\ell_0^2} \qquad (2.19)$$

with $z = x - iy$ being the electron position. The many-electron system is described by the Hamiltonian of the form:

$$\mathcal{H} = \sum_j \left[\frac{1}{2m_e}|-i\hbar\nabla_j - \frac{e}{c}\boldsymbol{A}_j|^2 + V(z_j)\right] + \sum_{j<k} \frac{e^2}{|z_j - z_k|} \qquad (2.20)$$

where z_j is the location of the jth electron and $V(z)$ is the potential generated by a uniform neutralizing background. It is easy to verify that [2.14],

$$\langle m|r^2|m\rangle = 2(m+1)\ell_0^2 \qquad (2.21)$$

which means that the area covered by a single electron in state $|m\rangle$, moving in its cyclotron orbit is proportional to m. This result might be considered as an indication of the relation between the interelectron spacing and the angular momentum. From (2.21), it is readily noticeable that the degeneracy of a Landau level N_s, is the upper bound to the quantum number m. This is seen by requiring that,

$$\pi \langle r^2 \rangle \leq A \qquad (2.22)$$

where A is the area of the system. One then obtains from the above two relations,

$$m \leq N_s - 1 \qquad (2.23)$$

where the Landau-level degeneracy N_s is defined in (2.7). From (2.19) and (2.23) we find that the state space of an electron in the lowest Landau level is spanned by $1, z, z^2, \ldots, z^{N_s-1}$ times the exponential factor $e^{-|z|^2/4\ell_0^2}$.

The *Jastrow-type* many-electron wave function proposed by Laughlin for $\nu = \frac{1}{m}$ state is

$$\psi_m = \prod_{j<k}^{N_e}(z_j - z_k)^m \prod_{j=1}^{N_e} e^{-|z_j|^2/4\ell_0^2}. \qquad (2.24)$$

For m being an odd integer, this wave function obeys Fermi statistics. The wave function is entirely made up out of states in the lowest Landau level. It is also an eigenstate of the angular momentum with eigenvalue $M = \frac{1}{2}N_e(N_e - 1)m$. The total angular momentum M is the degree of the polynomial (conservation of angular momentum). In order to gain more insight about the wave function, let us expand the first product in powers of z_1, keeping all other coordinates fixed. The highest power then would be $m(N_e - 1)$, which must be equal to $N_s - 1$ [see (2.23)]. For large N_e, we then obtain [see (2.7)],

$$m \cong \frac{A}{2\pi \ell_0^2 N_e} = \frac{1}{\nu}. \tag{2.25}$$

The parameter m is thus fixed by the density, and unlike in conventional Jastrow theory, we do not have a variational parameter in the trial wave function. For $m = 1$ (filled Landau level), the polynomial $\prod_{j<k}(z_j - z_k)$ is the Vandermonde determinant of order N_e. As $N_e \to \infty$, the particle density in this state tends to $1/(2\pi\ell_0^2)$ [2.15].

For $m > 1$, the wave function vanishes as a high power of the two electron separation, and thus tends to minimize the repulsive interaction energy. The probability distribution of the electron for ψ_m is given by

$$|\psi_m|^2 = e^{-\mathcal{H}_m} \tag{2.26}$$

with

$$\mathcal{H}_m = -2m \sum_{j<k} \ln|z_j - z_k| + \sum_j |z_j|^2 / 2\ell_0^2. \tag{2.27}$$

For a *charge neutral* two-dimensional classical plasma, the interaction is given [2.16] by

$$V(r) = -e^2 \sum_{j<k} \ln r_{jk} + \frac{1}{2}\pi \rho e^2 \sum_j r_j^2 \tag{2.28}$$

where the particles are interacting via a two-dimensional Coulomb (logarithmic) interaction with each other and with a uniform neutralizing background. From (2.27) and (2.28), it is clear that $\mathcal{H}_m$ is the Hamiltonian for a two-dimensional classical plasma with,

$$e^2 = 2m, \quad \rho_m = \frac{1}{2\pi\ell_0^2 m}. \tag{2.29}$$

Therefore, in order to achieve charge neutrality, the plasma particles spread out uniformly in a disk with particle density ρ_m, corresponding to a filling factor $\nu = \frac{1}{m}$, and m is an *odd* integer. The classical plasma provides a strong support that the Laughlin state is indeed a translationally invariant liquid [2.2,16].

The electrons however, in contrast to the classical plasma particles, interact via the *three-dimensional* Coulomb interaction, and the expectation value of the potential energy in a quantum state is given by

$$\frac{\langle V \rangle}{N_e} = \frac{1}{2} \int v(r)g(r)dr , \qquad (2.30)$$

where $g(r)$ is the two-particle radial distribution function, which will be calculated below by the classical plasma approach. Introducing the *ion-disk* radius [2.16] $R = \sqrt{2m}\ell_0$ and the dimensionless distance $x = r/R$, the energy per particle is given as

$$\langle E \rangle = \frac{1}{\sqrt{2m}} \frac{e^2}{\epsilon \ell_0} \int_0^\infty [g(x) - 1] \, dx. \qquad (2.31)$$

In (2.31), we have included the contribution from the neutralizing background. Because the wave function is for the lowest Landau level, the kinetic energy part is constant. The radial distribution function $g(x)$ was obtained by Laughlin using the hypernetted-chain (HNC)[1] theory, which is a well-established technique for dealing with the classical plasma [2.16,17] and quantum fluids [2.18].

The dimensionless plasma parameter $\Gamma = e^2/k_B T$ [2.16,17], is related to the present problem via, $\Gamma = 2m$. For $m = 1$, the *exact* result is available for the pair correlation function, $g(x) = 1 - e^{-x^2}$ and the energy, $E = -\sqrt{\pi/8}$ [2.19]. As *Laughlin* pointed out [2.2], $\Gamma = 2$ corresponds to a full Landau level where the total energy equals the HF energy. This correspondence explains the existence of an exact result for $\Gamma = 2$.

The two-dimensional Fourier transform is given by

$$\tilde{f}(q) = 2 \int_0^\infty f(x) J_0(qx) x \, dx \qquad (2.32)$$

where $q = kR$, and $J_0(x)$ is the zeroth-order Bessel function. For $m = 3, 5, 7, \ldots$ the distribution function is obtained from the following set of equations [the various functions are defined in Appendix B, see also (B.18)],

$$\begin{aligned} g(x) &\simeq \exp\left[N(x) - u(x)\right] \\ \tilde{N}(q) &= \left[\tilde{C}(q)\right]^2 / \left[1 - \tilde{C}(q)\right] \\ C(x) &= g(x) - 1 - N(x). \end{aligned} \qquad (2.33)$$

---

[1] An introduction to the HNC method is given in Appendix B.

with, $u(x) = -2m \ln x$. In order to handle the logarithmic interaction, the standard procedure is to separate the short-range and long-range part of the interactions as [2.17],

$$\begin{aligned} u^s(x) &= 2m K_0(Qx) \\ u^l(x) &= -2m \left[\ln x + K_0(Qx)\right] \end{aligned} \quad (2.34)$$

where K_0 is the modified Bessel function, and Q is the cutoff parameter of order unity. For small x, $u^s(x)$ reduces to the full Coulomb potential in two dimensions, and at large distances it decreases exponentially. Defining the short-range functions [2.20],

$$\begin{aligned} N^s(x) &= N(x) - u^l(x) \\ C^s(x) &= C(x) + u^l(x), \end{aligned} \quad (2.35)$$

one obtains the final set of equations

$$\begin{aligned} \widetilde{N}^s(q) &= \left\{\widetilde{C}^s(q)\left[\widetilde{C}^s(q) - \widetilde{u}^l(q)\right] - \widetilde{u}^l(q)\right\} / \left[1 - \widetilde{C}^s(q) + \widetilde{u}^l(q)\right] \\ \widetilde{u}^l(q) &= \frac{4mQ^2}{q^2(q^2 + Q^2)} \\ g(x) &= \exp\left[N^s(x) - u^s(x)\right] \\ C^s(x) &= g(x) - 1 - N^s(x) \end{aligned} \quad (2.36)$$

which are solved by a straightforward iteration scheme. As explained in Appendix B, the first step in the numerical iteration procedure could be to set, $\widetilde{N}^s(q) = 0$. The pair-correlation function is then trivially obtained with the known function $u^s(x)$. With that $g(x)$ one then obtains $C^s(x)$ and its Fourier transform $\widetilde{C}^s(q)$. Next step is to obtain $\widetilde{N}^s(q)$ from the new $\widetilde{C}^s(q)$ and the inverse Fourier transform $N^s(x)$ then provides the new $g(x)$. The process is repeated until convergence is achieved. The resulting $g(x)$ is plotted in Fig. 2.4 for $m = 3$ as a function of x, and shows characteristics of a *liquid* state. In Fig. 2.4, we also present the radial distribution function for a Wigner crystal state of the form [2.4]

$$\Psi_{\rm WC}(z_1, \ldots, z_n) = \sum_\sigma {\rm sgn}(\sigma)\, \phi_{j_1 k_1}[z_{\sigma(1)}] \cdots \phi_{j_n k_n}[z_{\sigma(n)}],$$

where σ is a permutation with ${\rm sgn}(\sigma)$ the sign and the orbitals are gaussians,

$$\phi_{jk}[z] = e^{-\frac{1}{4}\left|z_{jk}^{(0)}\right|^2} e^{\frac{1}{2} z^* z_{jk}^{(0)}} e^{\frac{1}{4}|z|^2},$$

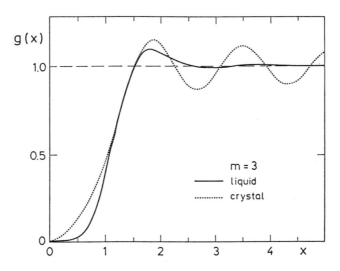

Fig. 2.4. Radial distribution function $g(x)$ as a function of the dimensionless interparticle separation $x = r/\sqrt{2m}\ell_0$, for the Laughlin state at $m = 3$ (solid line) and the WC state in the HF approximation (dashed line) at the same density

centered at hexagonal lattice sites $z_{jk}^{(0)}$,

$$z_{jk}^{(0)} = \sqrt{\frac{4\pi m}{\sqrt{3}}} \left[j + \left(\tfrac{1}{2} + \tfrac{1}{2}i\sqrt{3}\right) k \right].$$

For small x, $g(x)$ goes to zero as x^2 for the crystal state, while for the Laughlin state, it goes as x^{2m}.

The ground state energy for the $\nu = \tfrac{1}{3}$ state, obtained from the HNC $g(x)$ and (2.31) is $E(\nu = \tfrac{1}{3}) = -0.4056\, e^2/\epsilon\ell_0$. This result is quite close to the exact results obtained for finite electron systems in the earlier section. It should be remarked at this point that Laughlin used a modified HNC version where the so-called *elementary* diagrams (neglected in the above discussion) were approximately summed (modified HNC approach is briefly discussed in Appendix B) and the result was $-0.4156\, e^2/\epsilon\ell_0$. The ground state energies for various filling fractions can be fitted to the following approximate formula [2.2]:

$$E_{\rm pl}(m) \cong \frac{-0.814}{\sqrt{m}} \left[1 - \frac{0.230}{m^{0.64}} \right] \left(\frac{e^2}{\epsilon\ell_0} \right) \tag{2.37}$$

which is a smooth function of m.

Several authors have studied the important characteristics of Laughlin's wave function. *Halperin* [2.13] has shown that the m-fold vanishing of the wave function when two electrons come close, helps to minimize the potential energy of the system and also provides exceptional stability of the system at filling factors $\nu = \frac{1}{m}$. *Haldane* and *Rezayi* [2.9] found that for six particles on a sphere, the energy difference between the exact ground state and Laughlin's wave function is smaller than one part in 2000. They have also presented the Laughlin-Jastrow wave function for a geometry which satisfies periodic boundary conditions [2.21]. From this work they concluded that the success of Laughlin's wave function was because of the correct behavior as the particles approach each other and does not depend on the arguments based on the conservation of the angular momentum.

Utilizing the electron-hole symmetry in the lowest Landau level (see Sect. 2.1.), one obtains the Laughlin state for $\nu = 1 - \frac{1}{m}$, [2.3,22]. In the following, we present a brief discussion of the electron-hole symmetry. We follow closely the work of *Girvin* [2.22]. In the symmetric gauge, the Laughlin wave function is given by (2.24). As discussed in Appendix A, the exponential factor common to all wave functions can be eliminated by defining a Hilbert space of functions analytic in z with inner product,

$$(\psi, \phi) = \int d\mu(z)\, \psi^*(z)\phi(z) \tag{2.38}$$

with the measure

$$d\mu(z) = \frac{dx\,dy}{2\pi\ell_0^2} e^{-|z|^2/4\ell_0^2}. \tag{2.39}$$

Invoking particle-hole symmetry, the state with $\nu = 1 - \frac{1}{m}$ must have *holes* described by ψ_m in (2.24).

The properly normalized $(N+1)$-particle state for a filled Landau level is written as

$$\Phi_{N+1}(z_1, \ldots, z_{N+1}) = \frac{1}{Q_{N+1}} \prod_{j<k}^{N+1} (z_j - z_k)$$

$$Q_{N+1}^2 = (N+1)! \prod_{j=0}^{N} (2^j j!). \tag{2.40}$$

Inserting a hole in a single-particle state (2.19), we write

$$\theta_M(z_1, \ldots, z_N) = (N+1)^{\frac{1}{2}} \int d\mu(z_{N+1}) \phi_M^*(z_{N+1}) \Phi_{N+1}(z_1, \ldots, z_{N+1}). \tag{2.41}$$

It is to be noted that, Φ_{N+1} is a single Slater determinant,

$$\Phi_{N+1} = \frac{1}{Q_{N+1}} \sum_{\{P\}} (-1)^P \prod_{j=1}^{N+1} z_j^{P_j}. \tag{2.42}$$

In (2.42), P_j is the image of j under a permutation P of $N+1$ points $[0, 1, 2, \ldots, N]$. From (2.41) and (2.42), one obtains after integration,

$$\theta_M(z_1, \ldots, z_N) = (N+1)^{\frac{1}{2}} \frac{(2^M m!)^{\frac{1}{2}}}{Q_{N+1}} \sum_{\{P\}} (-1)^P \prod_{j=1}^{N} z_j^{P_j} \delta_{M, P_{N+1}}. \qquad (2.43)$$

The function θ_M describes an N-particle Slater determinant with every state ϕ occupied for $0 \le j \le N$ except the state ϕ_M, which remains empty. The norm is $(\theta_M, \theta_M) = 1$ as required. Equation (2.41) provides an exact procedure for injecting a hole into any particular state. One can, in fact, readily generalize (2.41) for the case of an arbitary number of holes. For example, the electron wave function,

$$\phi(z_1, \ldots, z_N) = \prod_{k=1}^{N} \int d\mu(z_{N+k}) \prod_{j<k}^{M} (z_{N+j}^* - z_{N+k}^*)^m \, \Phi_{N+M}(z_1, \ldots, z_{N+M}) \qquad (2.44)$$

would correspond to the *hole* wave function as the correlated state ψ_m of (2.24). Therefore, this state has the filling factor $\nu = 1 - \frac{1}{m}$ and for $N \to \infty$ corresponds to the *exact* particle-hole dual of the state with $\nu = \frac{1}{m}$.

Finally, in the case of repulsive interactions of vanishing range, *Trugman* and *Kivelson* [2.23] and independently *Pokrovskii* and *Talapov* [2.24] found that Laughlin's state ψ_m is the exact, nondegenerate ground state for $\nu = \frac{1}{m}$.

2.3 Spherical Geometry

An attractive alternative to the finite-size studies described in Sect. 2.1 is to consider the spherical geometry [2.8]. Here the electrons are confined on the surface of a sphere of radius R with a magnetic monopole at the center. The total magnetic flux, $\Phi = 4\pi R^2 B$ is required to be an integral multiple $N_s = 2S$ of the elementary flux quantum Φ_0. Therefore, the radius of the sphere is determined as

$$R = S^{\frac{1}{2}} \ell_0. \qquad (2.45)$$

For an electron of mass m on the surface of the sphere, the kinetic energy is written as

$$\mathcal{K} = |\Lambda|^2 / 2m = \frac{1}{2} \omega_c |\Lambda|^2 / \hbar S \qquad (2.46)$$

where $\Lambda = r \times [-i\hbar\nabla + e\mathbf{A}(r)]$ is the kinetic angular momentum; $\nabla \times \mathbf{A} = B\widehat{\Omega}$, with $\widehat{\Omega} = r/R$. The angular momentum Λ obeys the commutation relations,

$$[\Lambda_\alpha, \Lambda_\beta] = i\hbar\epsilon_{\alpha\beta\gamma}\left(\Lambda_\gamma - \hbar S\widehat{\Omega}_\gamma\right) \qquad (2.47)$$

where $\epsilon_{\alpha\beta\gamma}$ is the antisymmetric rank-three tensor. Let us now define the operators

$$\mathbf{L} = \Lambda + \hbar S\widehat{\Omega}, \qquad (2.48)$$

which are the generators of rotation with the commutation relations:

$$[L^\alpha, X^\beta] = i\hbar\epsilon_{\alpha\beta\gamma}X_\gamma \qquad (2.49)$$

where $X = L, \widehat{\Omega}$, or Λ.

As the angular momentum Λ has no component normal to the surface, $\Lambda \cdot \widehat{\Omega} = \widehat{\Omega} \cdot \Lambda = 0$, we have, $\mathbf{L} \cdot \widehat{\Omega} = \widehat{\Omega} \cdot \mathbf{L} = \hbar S$. As a result,

$$|\Lambda|^2 = |\mathbf{L}|^2 - \hbar^2 S^2 \qquad (2.50)$$

and the eigenvalues of $|\Lambda|^2$ are deduced from the usual angular momentum algebra, $|\Lambda|^2 = |\mathbf{L}|^2 - \hbar^2 S^2 = \hbar^2\{n(n+1) - S^2\}$, n being an integer. The gauge is chosen such that the vector potential is given by $\mathbf{A} = \left(\frac{\hbar S}{eR}\right)\widehat{\varphi}\cot\theta$.

For $n = S$, one obtains the lowest landau level with energy $\frac{1}{2}\hbar\omega_c$. The degeneracy of this level is in fact finite: There are $(2S+1)$ independent degenerate eigenfunctions. For complete occupation of these states, the electron density is given by

$$(2S+1)/4\pi R^2 \sim \frac{S}{2\pi R^2} = \frac{1}{2\pi\ell_0^2}$$

for large R, similar to what one obtains in a planar geometry.

In order to find the single-particle eigenstates of the Hamiltonian, Haldane [2.8] noted that these are most suitably represented in the space of spinors of rank $2S$. For that purpose he introduced the spinor variables, $u = \cos\left(\frac{1}{2}\theta\right)e^{i\varphi/2}$, $v = \sin\left(\frac{1}{2}\theta\right)e^{-i\varphi/2}$. The components of the operator $\mathbf{L}$ [see (2.48)] can now be written as

$$L^+ = \hbar u\frac{\partial}{\partial v}; \quad L_z = \frac{1}{2}\left(u\frac{\partial}{\partial u} - v\frac{\partial}{\partial v}\right)$$
$$L^- = \hbar v\frac{\partial}{\partial u} \qquad (2.51)$$

employing the standard raising and lowering operators $L^\pm = L_x \pm iL_y$. To satisfy, $\mathbf{L} \cdot \widehat{\Omega} = \hbar S$ one must have,

$$S = \frac{1}{2}\left(u\frac{\partial}{\partial u} + v\frac{\partial}{\partial v}\right).$$

For any homogeneous polynomial $\psi^{(S)}(u,v)$ of degree $2S$ the following relation holds,

$$|\boldsymbol{L}|^2 \psi^{(S)} = S(S+1)\hbar\psi^{(S)}. \tag{2.52}$$

One can now associate to each unit vector $\widehat{\boldsymbol{\Omega}}(\alpha,\beta)$, a homogeneous polynomial,

$$\psi^{(S)}_{(\alpha,\beta)}(u,v) = (\alpha^*u + \beta^*v)^{2S}, \; |\alpha|^2 + |\beta|^2 = 1$$

which also satisfies the eigenvalue equation

$$\left(\widehat{\boldsymbol{\Omega}}(\alpha,\beta)\cdot \boldsymbol{L}\right)\psi^{(S)}_{(\alpha,\beta)} = \hbar S \psi^{(S)}_{(\alpha,\beta)}. \tag{2.53}$$

A single electron may now represented as a spin S, the orientation of which indicates the point on the sphere about which the state is localized. It is interesting to note that the analogy between the spin wave functions and the electron wave functions in a magnetic field was first shown by *Peres* [2.25].

Turning our attention to the many-electron states, the Laughlin-type wave function for $\nu = \frac{1}{m}$ is then written as [2.8,9],

$$\psi_m = \prod_{j<k}(u_jv_k - v_ju_k)^m. \tag{2.54}$$

As a function of (u_j, v_j), wave functions are required to be polynomials of degree $2S$. Since the maximum degree in one variable is clearly $m(N_e - 1)$ we have $S = \frac{1}{2}m(N_e-1)$. On the other hand, we know that the Landau level is completely filled when all the $(2S+1)$ single-particle states are occupied. Therefore the product (2.54) corresponds to a filling fraction $\nu = \frac{1}{m}$. As any factor in the product (2.54) describes a pair of particles with total spin zero, one obtains an uniform electron density on the sphere. The wave function also has m-fold zeros when the coordinates of any two particles coincide. The wave function ψ_m is an eigenfunction of $\boldsymbol{L}^2$ with $L = 0$. This implies that $|\psi_m|^2$ is invariant under rotations of the sphere.

The numerical results for finite electron systems in this geometry were first obtained by *Haldane* and *Rezayi* [2.9] and later by *Fano* et al. [2.26]. The exact ground state at flux $2S = 3(N_e - 1)$, ($\nu = \frac{1}{3}$), was found to be isotropic with total angular momentum $L = 0$, and separated from the excited states at the same (N_e, S) by a gap. Their results for the ground state energy per particle are plotted in Fig. 2.5 for various values of N_e. The extrapolated result for $N_e \to \infty$ is $-0.415 \pm 0.005\, e^2/\epsilon\ell_0$. The energy values in this geometry are found to be more system-size dependent than those for the rectangular geometry.

Morf et al. [2.27] however noticed that, unlike in the disk geometry the areal electron density is size dependent:

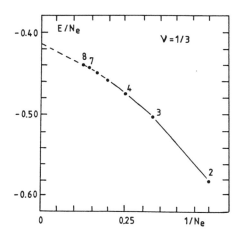

Fig. 2.5. Ground state energy per particle for finite electron systems in a spherical geometry as a function of electron number N_e [2.9]

$$\frac{N_e}{4\pi R^2} = \frac{N_e}{2\pi \ell_0^2} \frac{\nu}{N_e - 1}, \tag{2.55}$$

which leads to an energy which reflects this size dependence of the density. This size dependence in the energy can be removed by defining a modified magnetic length,

$$\ell_0' = (2S\nu/N_e)^{\frac{1}{2}} \ell_0 \tag{2.56}$$

such that $\nu/2\pi {\ell_0'}^2$ is independent of system size. For the interparticle separation, *Haldane* and *Rezayi* used the *chord* distance [2.9] between electrons i and j,

$$r(\Theta_{ij}) = 2R \sin \tfrac{1}{2}\Theta_{ij} \tag{2.57}$$

where the angle between position vectors r_i and r_j is written as

$$\Theta_{ij} = 2 \arcsin |u_i v_j - v_i u_j|. \tag{2.58}$$

One can in turn use the *great circle* distance,

$$r(\Theta_{ij}) = R\,\Theta_{ij}. \tag{2.59}$$

In the limit, $N_e \to \infty$, however, energies evaluated with these two definitions of distances converge to the same point (see Fig. 2.10).

We mentioned earlier that Haldane and Rezayi have studied the exactness of the Laughlin wave function. The concept of pseudopotentials first introduced by *Haldane* [2.8,9,28] is quite illuminating in this context. The pseudopotential parameters which enter into the interaction Hamiltonian

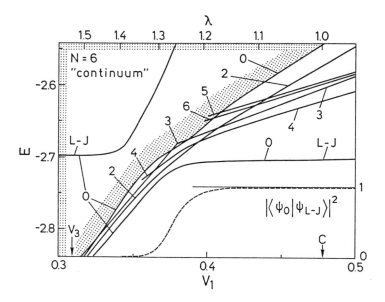

Fig. 2.6. Low-lying states at $\nu = \frac{1}{3}$ ($N_e = 6$, $2S = 15$) in the lowest Landau level, as V_1 is varied and $V_{m>1}$ are Coulombic. The Coulomb value of V_1 (marked as C) and V_3 are also shown. The values of angular momentum quantum number L are indicated in the figure. The projection of the Laughlin state on the ground state is shown. In the gapless regime ($\lambda > 1.25$), the Laughlin-Jastrow (LJ) state is seen to reappear as the *highest* $L = 0$ level [2.9]

are written

$$V_m = \frac{1}{(2\pi)^2} \int d\mathbf{r}\, V(r) \int d\mathbf{q}\, e^{i\mathbf{q}\cdot\mathbf{r}} e^{-\frac{1}{2}(q\ell_0)^2} \left[L_n(\tfrac{1}{2}q^2\ell_0^2) \right]^2 L_m(q^2\ell_0^2) \quad (2.60)$$

where $V(r)$ is the pair interaction potential, and $L_n(x)$ is the Laguerre polynomial. These are the correlation energies of pairs of particles with relative angular momentum $m \geq 0$. For small m, the parameters V_m describe the short-range part, while V_m with large m describe the long-range part of the interaction. One should note, however, that this description of the interaction is not a local decomposition in real space. The Hamiltonian for the lowest Landau level can be written as $\mathcal{H}_0 + \lambda \mathcal{H}_1$, where $\lambda = 0$ represents the truncated pseudopotential where only $V_1 \neq 0$. For $\lambda = 1$, we have the full Coulomb interaction in the lowest Landau level.

In Fig. 2.6, we reproduce the results of [2.9] for the effect of variation of the short-range component V_1, while the other pseudopotential parameters ($V_{m>1}$) are kept at their Coulomb values. In the limit, $V_1 \to +\infty$, the Laughlin state is the only state for which the matrix element $\langle \psi | \mathcal{H} | \psi \rangle$ is independent of V_1, and becomes the exact ground state. The other states have energies which are proportional to V_1 in this limit, and there is a large energy gap. As we lower V_1, the gap persists for $\lambda \leq 1.25$, and in this regime, the Laughlin state has an almost 100% projection on the true ground state. This region also includes the ground state for the pure lowest Landau level Coulomb interaction, which is therefore well described by the Laughlin state.

At $\nu = \frac{1}{3}$, the Coulomb potential therefore has the stability limit of $\lambda \approx 1.25$, where there is a first order transition to a gapless state, which was found to be compressible. The Laughlin state is then found as an *excited* state in the gapless regime.

2.4 Monte Carlo Results

As we have seen in the earlier sections, Laughlin's wave function together with the classical plasma theory has provided quite a reliable description of the ground state and is in good agreement with the exact evaluation of the ground state energies for finite electron systems. There have also been very accurate Monte Carlo studies to confirm these results on quantitative grounds.

Levesque et al., [2.5] obtained the ground state energy of the Laughlin wave function by evaluating the pair correlation functions for about 256

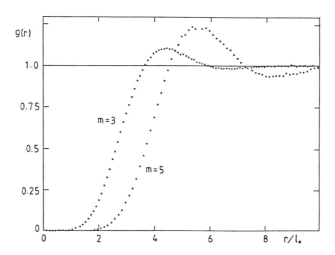

Fig. 2.7. Pair correlation function $g(r)$ for $\nu = \frac{1}{3}$ and $\frac{1}{5}$, obtained from Monte Carlo calculations [2.5]

particles, using the method described by *Caillol* et al. [2.16]. The results for $\nu = \frac{1}{3}$ and $\frac{1}{5}$ are shown in Fig. 2.7. For the ground state energy they obtained $E(\nu = \frac{1}{3}) = -0.410 \pm 0.0001$ and $E(\nu = \frac{1}{5}) = -0.3277 \pm 0.0002$ (in units of $e^2/\epsilon \ell_0$). A comparison of these almost *exact* results with those obtained via the exact diagonalization for few electron systems and the classical plasma approach described in Sects. 2.1 and 2.2 respectively, should convince the reader of the efficiency of these methods in describing the electron system. This credibility is quite essential, as we proceed to discuss other quantities of interest with these methods in the following chapters.

As a function of filling factor, Levesque et al. obtained the following relation for the energy by fitting their results and the exact result for $m = 1$:

$$E \simeq -0.782133\sqrt{\nu}\left(1 - 0.211\,\nu^{0.74} + 0.012\,\nu^{1.7}\right). \qquad (2.61)$$

In the disk geometry, *Morf* and *Halperin* [2.6] performed a Monte Carlo evaluation of the ground state energy for up to 144 electrons. The results are shown in Fig. 2.8. The results for the energy (per particle) are fitted accurately by the polynomial,

$$E \approx -0.4101 + 0.06006/\sqrt{N_e} - 0.0423/N_e. \qquad (2.62)$$

Using fits by first-, second-, and third-order polynomials in $N_e^{-\frac{1}{2}}$ results are obtained for the thermodynamic limit as $-0.4092, -0.4101$ and -0.4099

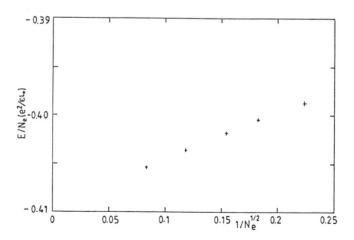

Fig. 2.8. Energy per electron for the $\nu = \frac{1}{3}$ state. Results are obtained for Laughlin's wave function via Monte Carlo calculations in disk geometry and plotted versus $N_e^{-\frac{1}{2}}$ for systems with N_e=20, 30, 42, 72, and 144 electrons. The length of the vertical bars indicate the standard deviation [2.6]

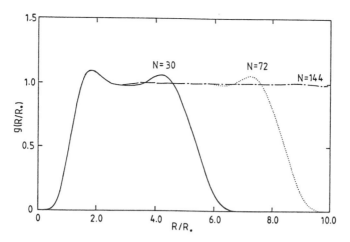

Fig. 2.9. Monte Carlo results for the radial distribution function $g(r)$ for the Laughlin state at $\nu = \frac{1}{3}$. Here R_0 is the ion-disk radius. Results are based on calculations with N_e=30, 72, and 144 electrons [2.6]

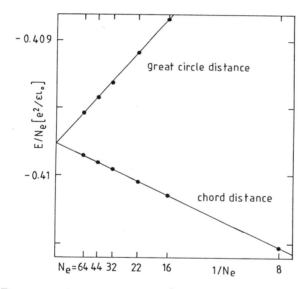

Fig. 2.10. Energy per electron for the $\nu = \frac{1}{3}$ state with Laughlin's wave function in a spherical geometry, as a function of $1/N_e$. The Coulomb energy is evaluated using the *great circle* and *chord* distance definitions [2.7]

(in units of $e^2/\epsilon\ell_0$) respectively. The radial distribution function $g(r)$ obtained from the Monte Carlo calculations of the Laughlin state at $\nu = \frac{1}{3}$ is shown in Fig. 2.9. There is little size dependence seen for small interparticle separation. From the distribution function thus obtained, Morf and Halperin also evaluated the Coulomb energy per particle. The results are, $-0.4096\, e^2/\epsilon\ell_0$ for $N_e=42$ and 72, and $-0.4097\, e^2/\epsilon\ell_0$ for a system with 144 particles. Their estimate in the thermodynamic limit for the energy per particle $\approx -0.410 \pm 0.001\, e^2/\epsilon\ell_0$ is in good agreement with the results of Levesque et al. described earlier in this section.

Finally, Monte Carlo results for the spherical geometry are also available [2.7] and are shown in Fig. 2.10. A least-squares fit to these results leads to

$$E \approx -0.40973 - 0.0072/N_e + 0.009/N_e^2$$

for the chord distance definition of the interparticle separation, and

$$E \approx -0.40975 + 0.012/N_e + 0.048/N_e^2$$

for the great circle distance definition. The finite-size corrections are obviously very small.

2.5 Reversed Spins in the Ground State

In our discussions so far, it has been assumed that because of the strong magnetic fields involved, only one spin state is present. *Halperin* first pointed out [2.13] that in GaAs, the electron g-factor is one-quarter of the free electron value [2.29]. Therefore, the Zeeman energy is approximately sixty times smaller than the cyclotron energy. It might then be possible to have some electrons with *reversed* spins when the magnetic field is not too large. In the case when one half of the electrons have spins antiparallel to the field, Halperin constructed a simple Laughlin-type state of the form

$$\psi = \prod_{i<j}(z_i - z_j)^m \prod_{\alpha<\beta}(z_\alpha - z_\beta)^m \prod_{i,\alpha}(z_i - z_\alpha)^n \prod_i e^{-|z_i|^2/4\ell_0^2} \prod_\alpha e^{-|z_\alpha|^2/4\ell_0^2}$$

(2.63)

where Greek and Roman indices correspond to spin-up and spin-down electrons respectively. Using the classical plasma approach (2.26–29), the filling factor is given by

$$\nu = \frac{2}{(m+n)}.$$

The ground state energy for the two-spin state is obtained by considering the spin-up and spin-down electrons as two different species of particles.

The total interaction energy is then obtained from

$$E_{\text{int}} = \frac{1}{2\sqrt{m+n}} \frac{e^2}{\epsilon \ell_0} \int_0^\infty [g_{11}(x) + g_{12}(x) - 2] \, dx \tag{2.64}$$

where $g_{\alpha\beta}$ are the partial pair-correlation functions.

For $\nu = \frac{2}{5}$, a generalized two-component HNC approach was employed by *Chakraborty* and *Zhang* [2.30] to obtain $g_{\alpha\beta}$ (for details see Appendix B) and the ground state energy was obtained as, $E_{\text{unpol}} = -0.434 \, e^2/\epsilon\ell_0$. For a spin-polarized state at this filling, a Laughlin-type state does not exist. In order to compare the unpolarized state with a spin polarized state for $\nu = \frac{2}{5}$, the above authors first performed a Laughlin-type calculation for $\nu = \frac{2}{5}$, which was then corrected with Halperin's hierarchial approach described in Sect. 3.7. The resulting energy for the fully spin-polarized state was, $E_{\text{pol}} = -0.429 \, e^2/\epsilon\ell_0$. Since then, several different calculations have been reported in the literature [2.6,27] and the result quoted above is close to the Monte Carlo data, $E_{\text{pol}} \approx -0.4303 \pm 0.003 \, e^2/\epsilon\ell_0$. Comparing these results, it can be concluded that, in the absence of Zeeman energy, a spin-unpolarized state for $\nu = \frac{2}{5}$ would be energetically favored. The Zeeman energy (per particle) is evaluated from $E_{\text{Zeeman}} = (1 - 2p)g\mu_B B s$, where p is the ratio of the number of spins parallel to the field to the total number of spins, $\mu_B = e\hbar/2mc$ the Bohr magneton, and $s = \frac{1}{2}$. For GaAs with all spins parallel to the field, $E_{\text{Zeeman}} = -0.011 \, e^2/\epsilon\ell_0$ for a magnetic field of 10 T, the Landé g-factor, $g \simeq 0.52$, and $\epsilon \simeq 13$.

In a later investigation [2.31], a systematic study of spin reversal in various filling fractions was attempted. The finite-size calculations described in Sect. 2.1, are particularly suitable for this work, since one need not construct trial wave functions for all the spin states to be studied and comparison is possible for various spin polarizations at different filling factors evaluated within the same numerical accuracy. The method of *Yoshioka* et al., [2.1] can be generalized for various spin polarizations in a straightforward manner. The energy spectrum for the Hamiltonian is classified in terms of the total spin S and its z-component S_z. For a given S, the spectrum is identical for different values of S_z.

In Table 2.1, we present the results for a finite electron system for the case of a polarized state ($S = 2$), a partly polarized state ($S = 1$) and the unpolarized state ($S = 0$). As seen in Table 2.1, except for $\nu = \frac{1}{3}$, the unpolarized states are energetically favored, as compared to the fully polarized state. In fact, for all the filling fractions considered in this work, the lowest energy is found to correspond to the case where the partial filling factor for each spin state has an *odd* denominator. It is interesting to note that, $\nu = \frac{4}{9}$ has a partial filling factor of $\frac{1}{3}$ for $S = 1$. Based on the fact

Table 2.1. Potential energy (per particle) for the four-electron system at various filling factors with the polarized ($S = 2$), partly polarized ($S = 1$), and the unpolarized ($S = 0$) electron spins, and the partial filling fractions $\nu_\uparrow$ and $\nu_\downarrow$ of the ground states. The Zeeman energy is not included in the energy values. The unit of energy is $e^2/\epsilon\ell_0$.

	Potential energy			Ground state		
ν	$S = 2$	$S = 1$	$S = 0$		$\nu_\uparrow$	$\nu_\downarrow$
$\frac{1}{3}$	-0.4152	-0.4120	-0.4135	Polarized	$\frac{1}{3}$	0
$\frac{2}{7}$	-0.3870	-0.3868	-0.3884	Unpolarized	$\frac{1}{7}$	$\frac{1}{7}$
$\frac{2}{5}$	-0.4403	-0.4410	-0.4464	Unpolarized	$\frac{1}{5}$	$\frac{1}{5}$
$\frac{4}{13}$	-0.3975	-0.3997	-0.3970	Partially polarized	$\frac{3}{13}$	$\frac{1}{13}$
$\frac{4}{11}$	-0.4219	-0.4278	-0.4241	Partially polarized	$\frac{3}{11}$	$\frac{1}{11}$
$\frac{4}{9}$	-0.4528	-0.4600	-0.4554	Partially polarized	$\frac{1}{3}$	$\frac{1}{9}$

that the filling factor $\nu = \frac{1}{3}$ is an experimentally observed stable state, it is quite tempting to predict that $\nu = \frac{4}{9}$ should be a stable state with partially polarized electron spins.

Including the Zeeman energy as estimated above, it was found that, except for $\nu = \frac{4}{11}$ and $\frac{4}{9}$, the spin-polarized state was energetically favored for all the other filling fractions considered in this work. For these two filling fractions, even in the presence of Zeeman energy, a partly spin-polarized state was found to have the lowest energy. The fraction $\nu = \frac{4}{9}$ has been observed recently by *Störmer* and his collaborators [2.32].

One obvious limitation of the above work was the small system size. Inclusion of the spin degree of freedom causes the Hamiltonian matrix to increase very rapidly. However, for the spin polarized state at $\nu = \frac{4}{9}$, the numerical calculations of *Yoshioka* [2.12] showed that the energy for eight electrons is only ~ 0.002 higher than that for four electrons. If this trend persists even for other spin polarizations, the conclusions would not be severely altered. However, no larger system calculation exists yet, to justify this expectation.

For $\nu = \frac{1}{3}$, the spin state $S = 2$ is found to be energetically favored compared with the other spin states, even in the absence of Zeeman energy. This result is quite supportive of Laughlin's state at $\nu = \frac{1}{3}$, which is fully antisymmetric. For other filling fractions, the possibility exists that the energy could be lowered by introducing the spin degree of freedom. In

Sect. 3.3, we shall see that spin reversal plays a more direct role in the elementary excitations.

For multivalley semiconductors like Si, it has been predicted [2.33] that there should be spontaneous valley polarization at $\frac{1}{3}$ and $\frac{1}{5}$. The valley degeneracy can be mapped on the spin systems (without having to add the Zeeman energy) and the results of Table 2.1 have been shown to hold in this case [2.34]. It has been shown that a gapless Goldstone mode exists for $\frac{1}{3}$ and $\frac{1}{5}$ states. These *valley waves* are analogous to the spin waves described in Sect. 4.3; see also [2.35].

2.6 Finite Thickness Correction

In our discussions of the electron system thus far, we have ignored the finite spread of the electron wave function perpendicular to the two-dimensional plane. In the real systems, the wave function of an electron has a finite spread perpendicular to the two-dimensional plane, in the z-direction. It is well known [2.36–39] that inclusion of the finite thickness correction effectively softens the short-range divergence of the bare Coulomb interaction, when the interelectron spacing is comparable with the inversion layer width.

The most common form assumed for the charge distribution normal to the plane is the Fang-Howard variational wave function, first proposed for the inversion layer. It is written as

$$g(z) = \frac{1}{2} b^3 z^2 \exp(-bz) \qquad (2.65)$$

where the effect of only the lowest subband is considered. The variational parameter is given by

$$b = \left[33\pi m^* e^2 n / 2\epsilon \hbar^2 \right]^{\frac{1}{3}} \qquad (2.66)$$

where $m^* \sim 0.067 m_e$ is the effective mass for GaAs, n is the electron density fixed by the Landau level filling ν. The depletion layer electron density is negligible as compared to the electron density and is not included in (2.66). For $n = 10^{11} \text{cm}^{-2}$, one obtains $b^{-1} \simeq 58\text{Å}$. On the other hand, the magnetic length can be expressed in terms of the magnetic field as, $\ell_0 = 256.6 B^{-\frac{1}{2}}$ with B in Tesla. For $B = 20\text{T}$, $\ell_0 \sim 57\text{Å}$ — the spread of the wave function is expected to have a substantial effect on the physical quantities of interest.

The effective electron-electron interaction is then written as [2.39],

$$V(r) = \frac{e^2}{\epsilon \ell_0} \int dz_1 dz_2 \, g(z_1)g(z_2)\left[r^2 + (z_1-z_2)^2\right]^{-\frac{1}{2}}. \quad (2.67)$$

Using the well-known Fourier transform result

$$F(q) \equiv \frac{q}{2\pi} \int \frac{dr \, e^{i q \cdot r}}{\left[(z_1-z_2)^2 + r^2\right]^{\frac{1}{2}}} \quad (2.68)$$
$$= \exp\left[-q|z_1-z_2|\right]$$

and $g(z)$ from (2.65), the interaction term is rewritten as

$$V(r) = \left[\frac{1}{2}b^3\right]^2 \left(\frac{e^2}{\epsilon \ell_0}\right) \int_0^\infty dq J_0(qr) \int dz_1 dz_2 \, z_1^2 z_2^2 e^{-b(z_1+z_2)} e^{-q|z_1-z_2|}. \quad (2.69)$$

After some algebra the final result is then obtained as

$$V(r) = \left(\frac{e^2}{\epsilon \ell_0}\right) \int_0^\infty dq F(q) J_0(qr) \quad (2.70)$$

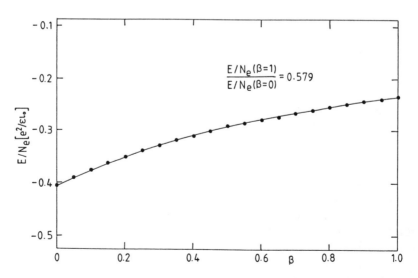

Fig. 2.11. Ground state energy (per particle) as a function of the dimensionless thickness parameter $\beta = (b\ell_0)^{-1}$ for the Laughlin wave function

with

$$F(q) = \left[1 + \frac{9}{8}\frac{q}{b} + \frac{3}{8}\left(\frac{q}{b}\right)^2\right]\left(1 + \frac{q}{b}\right)^{-3}. \tag{2.71}$$

For large r, one has the usual $1/r$ behavior, and for small r the $\ln r$ behavior is obtained.

In the case of Laughlin's wave function for $\nu = \frac{1}{3}$, *MacDonald* and *Aers* [2.38] and *Chakraborty* [2.40] obtained substantial reduction of the ground state energy as a function of $\beta = (b\ell_0)^{-1}$, as shown in Fig. 2.11. The reduction of the energy is a direct consequence of the softening of the Coulomb repulsion. For the excitation energy gap, finite-system calculations were done by *Zhang* and *Das Sarma* [2.39] and *Chakraborty* et al. [2.41] and for the Laughlin quasihole by *Chakraborty* [2.40]. These results will be discussed in Chap. 3.

2.7 Liquid-Solid Transition

It is perhaps evident from the above discussions that the ground state for $\nu = \frac{1}{m}$ and $1 - \frac{1}{m}$ (and other higher order filling fractions to be described in Sect. 3.7) is the quantum fluid state, and is lower in energy than the crystal state. On the other hand, at low density, i.e. for very small ν, the ground state is expected to be a Wigner crystal. It should be noted that the two-dimensional one-component classical plasma has a crystallization transition which occurs at $\Gamma \simeq 140$, i.e. for $m = 70$ [2.16]. Therefore, it would be interesting to estimate the filling factor at which the liquid to solid transition takes place for the present quantum system.

Most computations of the crystal energy have been within the HF approximation [2.1,5,10,42]. Comparing the accurate Monte Carlo results for the Laughlin liquid state (discussed in Sect. 2.4) with the HF crystal energies as a function of filling factor, the critical filling was found to occur at $\nu_c \sim \frac{1}{10}$ [2.5]. In order to improve the calculation for the crystal state, *Lam* and *Girvin* [2.43] used a variational wave function for a *correlated* Wigner crystal and minimizing the energy, obtained a relation valid for $\nu \leq \frac{1}{2}$:

$$E_{\text{WC}}^{\text{c}} = -0.782133\,\nu^{\frac{1}{2}} + 0.2410\,\nu^{\frac{3}{2}} + 0.16\,\nu^{\frac{5}{2}}. \tag{2.72}$$

Comparing the energies with the Monte Carlo results of [2.5], Lam and Girvin estimated the crossover point to be $\frac{1}{\nu_c} = 6.5 \pm 0.5$. This result is, of course, an estimate for an *ideal* system. In a real system, the finite-thickness correction (Sect. 2.6) and Landau-level mixing would influence this estimate somewhat. Experiments supporting the above prediction have been performed by *Mendez* et al. [2.44], who indicated that $\frac{1}{\nu_c}$ may be seven or larger. This conclusion was based on the absence of minima of ρ_{xx} at $\nu = \frac{1}{7}$ and $\frac{1}{9}$ in a low-density sample where a weak structure could still be observed at $\nu = \frac{1}{5}$.

Chang et al. [2.45] however argued that the weakness of the $\frac{1}{5}$ is most likely a result of localization, and the gap is reduced significantly due to disorder. They argue that a similar reduction in the energy gap also occurs for $\frac{1}{7}$, $\frac{1}{9}$, $\frac{1}{11}$ etc. The critical filling factor at which the liquid-solid transition takes place is still a very interesting open question.

3. Elementary Excitations

One important result of Laughlin's theory was the observation that the elementary charged excitations in a stable state $\nu = \frac{1}{m}$ are the quasiparticles and quasiholes with *fractional* electron charge of $\pm \frac{e}{m}$. If one electron is added to the system, it amounts to adding m elementary excitations and hence, the discontinuity in slope of the energy curve can be written as

$$\left.\frac{\partial E}{\partial N_e}\right|_{\nu_+} - \left.\frac{\partial E}{\partial N_e}\right|_{\nu_-} = m(\widetilde{\varepsilon}_p + \widetilde{\varepsilon}_h) = mE_g \qquad (3.1)$$

where E_g is the energy required to create one quasiparticle (with energy $\widetilde{\varepsilon}_p$) and one quasihole (with energy $\widetilde{\varepsilon}_h$) well separated from each other. The pinning of the density to $\nu = \frac{1}{3}$ suggests that there will be no low frequency phonon-type of excitations at long wavelengths.

In the *absence* of impurities, there will be no dissipation associated with the flow of current leading to vanishing longitudinal conductivity. In the presence of a weak random potential which does not close up the energy gap, there will still be no dissipation. The uniform electric field E causes the electrons to drift. In a frame of reference moving with the drift velocity $v_D = (E \times B)c/B^2$, the electric field is zero and so there is no electric current. The impurities however move relative to the frame with velocity $-v_D$. Because of the energy gap, at low temperatures the time dependent impurity potential will not generate any excitations and hence no dissipation of energy.

When the filling factor ν is slightly shifted from the stable state $\frac{1}{m}$, with m being an odd integer, the ground state of the system is expected to consist of a small density of quasiparticles or quasiholes, with charge $\pm \frac{e}{m}$ and Coulomb interactions. In the presence of impurities, these quasiparticles or quasiholes are expected, for low concentrations, to be trapped in potential fluctuations.

In the following sections, we describe how the quasiparticle and quasihole states are constructed and various methods are applied to evaluate the energy gap. Several experiments have been reported so far on the energy gap, which is related to the activation energy measured from thermal activation of ρ_{xx}. Finite-size calculations are also useful in obtaining E_g and there have been accurate results from Monte Carlo calculations. The following discussion of Laughlin's quasiparticle and quasihole states is based on the articles by *Laughlin* [3.1,2], *Halperin* [3.3–6] and *Chakraborty* [3.7].

3.1 Quasiholes and Quasiparticles

As we mentioned earlier, Laughlin's ground state wave function has m-fold zeros when two particles come close to each other. *Halperin* [3.3] first pointed out that if, in Laughlin's state (2.24), we fix the positions of all electrons except one (say z_1) and move that electron around a closed loop, avoiding other electrons in the system, the phase of the wave function changes by $\Delta\phi \sim 2\pi \frac{\Phi}{\Phi_0}$ for large area A. [In fact, if the test electron goes around say the second electron, we clearly have $(z_1 - z_2)^m = r^m e^{im\theta}$, and when θ is changed by -2π, the phase of the wave function changes by $-2\pi m$.] Recalling that, $\frac{\Phi}{\Phi_0} = N_e m$ [see (2.8) and (2.25)], we see that, for each electron within area A, the wave function must have $N_e m$ zeros as a function of the particle coordinates. This amounts to saying, we have one zero per flux quantum or m zeros per particle. Defining a *vortex* as a point where the wave function is zero in such a way that the phase changes by -2π for counterclockwise rotation, Halperin pointed out that Laughlin's wave function has precisely m vortices at each electron position and no other *wasted* vortices elsewhere in the sample.

Quasiholes: For densities slightly different from $\nu = \frac{1}{m}$, we cannot construct a wave function with exactly m vortices tied to each electron. In order to have electron density slightly *less* than $\frac{1}{m}$, either we add a few extra vortices not tied to electron positions, or have some electrons with more than m vortices. Laughlin considered the first choice, which is easier to realize. His wave function is written as [3.1] (in units where $\ell_0 = 1$)

$$\psi_m^{(-)} = e^{-\frac{1}{4}\sum_l |z_l|^2} \prod_j (z_j - z_0) \prod_{j<k} (z_j - z_k)^m \quad (3.2)$$

where $z_0 = x_0 - iy_0$. This wave function has a simple zero at $z_j = z_0$ for any j, as well as m-fold zeros at each point where $z_j = z_k$, for $k \neq j$. Writing

$$|\psi_m^{(-)}|^2 = e^{-\mathcal{H}_m^{(-)}}, \quad (3.3)$$

we obtain

$$\mathcal{H}_m^{(-)} = \mathcal{H}_m + 2\sum_j \ln |z_j - z_0|, \quad (3.4)$$

which is just the Hamiltonian of a classical one-component plasma in the presence of an extra repulsive *phantom* point charge at point z_0, whose strength is less by a factor $\frac{1}{m}$ than the charges in the plasma. The plasma will neutralize this phantom by a *deficit* of $\frac{1}{m}$ charge near z_0, while elsewhere in the interior of the system, the charge density will not be changed. However, the real three-dimensional electric charge is carried by the electrons and by the uniform positive background, and *not* by the phantom. As the electron charge density cancels the uniform background, a real net charge $-\frac{e}{m}$ is accumulated in the vicinity of z_0. The wave function (3.2) therefore describes a *quasihole* at point z_0.

The quasihole creation energy is calculated as follows: The HNC equations for a two-component system are written as [see (2.32–35)][1]

$$g_{\alpha\beta}(x) = \exp\left[N_{\alpha\beta}(x) - u_{\alpha\beta}(x)\right]$$
$$\widetilde{N}_{\alpha\beta}(q) = \sum_{\gamma=1,2} \rho_\alpha \widetilde{C}_{\gamma\alpha}(q) \left[\widetilde{C}_{\gamma\beta}(q) + \widetilde{N}_{\gamma\beta}(q)\right] \quad (3.5)$$
$$C_{\alpha\beta}(x) = g_{\alpha\beta}(x) - 1 - N_{\alpha\beta}(x).$$

Here $u_{\alpha\beta}(x) = 2[1 - (1-m)\delta_{\alpha\beta}]\ln x$, the indices α, β run over two types of particles, and ρ_α is the number density of species α. A single phantom particle in an otherwise uniform system can be considered as the *impurity limit* of the two-component HNC theory. In this limit, defining $c = \rho_2/\rho$, $\rho = \rho_1 + \rho_2$, one considers the case $c \to 0$. The equations (3.5) then decouple [3.7], and

$$\widetilde{N}_{\alpha\beta}(q) = \rho_\alpha \widetilde{C}_{\alpha\alpha}(q)\widetilde{C}_{\alpha\beta}(q) / \left[1 - \rho_\alpha \widetilde{C}_{\alpha\alpha}(q)\right]. \quad (3.6)$$

The other two functions $g_{\alpha\beta}(x)$ and $C_{\alpha\beta}$ are given as above. Defining the short-range functions as in (2.34), the final form of the HNC equations in the presence of an impurity are

$$\widetilde{N}^s_{\alpha\beta}(q) = \left[\widetilde{C}^s_{\alpha\beta}(q)\widetilde{C}_{\alpha\alpha}(q) - \widetilde{u}^1_{\alpha\alpha}(q)\right]$$
$$C^s_{\alpha\beta}(x) = g_{\alpha\beta}(x) - 1 - N^s_{\alpha\beta}(x) \quad (3.7)$$
$$g_{\alpha\beta}(x) = \exp\left[N^s_{\alpha\beta}(x) - u^s_{\alpha\beta}(x)\right].$$

These equations are solved for $g_{11}(x)$ and $g_{12}(x)$ by an iterative scheme. One first solves the equations for the background plasma [thus obtaining $C_{11}(x)$

---

[1] See Appendix B for the explanation of the various functions.

and $g_{11}(x)$]. The Fourier transform $\widetilde{C}_{11}(q)$ is then used in the impurity equations to solve for $g_{12}(x)$ and $C_{12}(x)$.

The quasihole creation energy is obtained from

$$\widetilde{\varepsilon}_{\rm h} = \frac{1}{\sqrt{2m}} \int_0^\infty dx\, \delta g_{11}(x) \qquad (3.8)$$

where δg_{11} corresponds to the charge in the background function $g_{11}(x)$ due to the presence of the impurity particle. This is calculated [3.7] from

$$\delta g_{11}(x) = \lim_{c \to 0} \frac{dg_{11}(x)}{dc}. \qquad (3.9)$$

From (3.7) one can readily obtain the final set of equations:

$$\begin{aligned}
\delta g_{11}(x) &= g_{11}(x)\, \delta N_{11}^{\rm s}(x) \\
\delta \widetilde{N}_{11}^{\rm s}(q) &= \left\{ \delta \widetilde{C}_{11}^{\rm s}(q) \left[\widetilde{h}_{11}(q) + \widetilde{C}_{11}(q)\right] + \widetilde{C}_{12}(q) h_{12}(q) \right\} \Big/ [1 - \widetilde{C}_{11}(q)] \\
\delta C_{11}^{\rm s}(x) &= \delta g_{11}(x) - \delta N_{11}^{\rm s}(x).
\end{aligned} \qquad (3.10)$$

In Fig. 3.1, we have plotted $\delta g_{11}(x)$ as a function of x for $m = 3$. The quasihole creation energy is obtained to be $\widetilde{\varepsilon}_{\rm h} = 0.0276\, e^2/\epsilon \ell_0$ for $m = 3$ and $0.0088\, e^2/\epsilon \ell_0$ for $m = 5$.

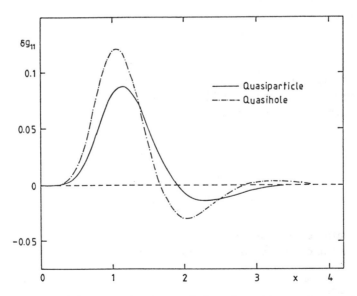

Fig. 3.1. Plot of $\delta g_{11}(x)$ versus x for the quasihole and quasielectron for $m = 3$

Laughlin [3.2] derived the quasihole creation energy in a slightly different way. He started with the two-component HNC equations (3.5). The particle densities are $\rho_1 = 1$ and $\rho_2 = \frac{1}{N_e}$. Using ρ_2 as the small parameter, he then solved the two-component HNC equations perturbatively. To zeroth order, the equations decouple and we get, $g_{11}(x) = g(x)$, which is the ground state function. More explicitly, one obtains[2]

$$\begin{aligned}\tilde{h}_{11}(q) &= \tilde{C}_{11}(q) + 2\tilde{h}_{11}(q)\tilde{C}_{11}(q) \\ \tilde{h}_{12}(q) &= [1 + 2\tilde{h}_{11}(q)]\tilde{C}_{12}(q).\end{aligned} \quad (3.11)$$

To first order, one would get the relation

$$h_{11} + \delta h_{11} = C_{11} + \delta C_{11} + 2(h_{11} + \delta C_{11}) + \frac{2}{N_e}h_{12}C_{12} + \left[O\left(\frac{1}{N_e}\right)^2\right]. \quad (3.12)$$

From the above two equations we readily obtain,

$$\delta h_{11} = [1 + 2h_{11}]^2 \delta C_{11} + \frac{2}{N_e}h_{12}^2. \quad (3.13)$$

The quasihole creation energy is then calculated from,

$$\tilde{\varepsilon}_h = \frac{N_e}{\sqrt{2m}} \int_0^\infty dx\, \delta h_{11} \quad (3.14)$$

to be $\tilde{\varepsilon}_h = 0.026\, e^2/\epsilon \ell_0$ for $m = 3$ and $0.008\, e^2/\epsilon \ell_0$ for $m = 5$. Recently, Fertig and Halperin [3.5] studied a generalization of the HNC approximation and calculated the three-body correlation functions g_{112} for a classical plasma with two particle species. Their result for the quasihole creation energy is $\tilde{\varepsilon}_h = 0.028\, e^2/\epsilon \ell_0$.

Introducing the finite thickness correction, (see Sect. 2.6), the quasihole creation energy has been calculated by one of us [3.8] and is plotted in Fig. 3.2, as a function of the dimensionless thickness parameter $\beta = (b\ell_0)^{-1}$.

Finally, because the charge accumulated around the phantom is determined by the long-range behavior of the interaction, the *particle excess* can be calculated as [3.4]

$$X = 2\int_0^\infty [g_{12}(x) - 1]\, x\, dx. \quad (3.15)$$

---

[2] One of us (T. C.) would like to acknowledge very helpful discussions about this derivation with B. I. Halperin and H. Fertig.

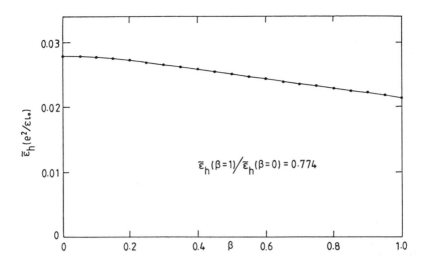

Fig. 3.2. The proper excitation energy for a Laughlin quasihole as a function of the finite thickness parameter β

Using our result for $g_{12}(x)$ from the HNC theory (Fig. 3.3) we obtain the expected result [3.8], $X = -0.328$ or $\sim -\frac{1}{3}$.

Quasiparticles: In the case when the electron density is slightly *higher* than the stable $\frac{1}{m}$ state, the choice of the wave function is not so clear. In this case, one clearly needs a state with one flux quantum (or equivalently, one zero of the wave function) missing. The wave function proposed by Laughlin for the quasiparticle state is written

$$\psi = \prod_{j=1}^{N_e}\left[e^{-|z_j|^2/4\ell_0^2}\left(2\ell_0^2\frac{\partial}{\partial z_j} - z_0^*\right)\right]\prod_{l<k}(z_l - z_k)^m. \tag{3.16}$$

In this case, the square of the wave function is not directly interpretable as the distribution in a classical statistical mechanics problem. However, *Laughlin* [3.2] has provided a means to calculate the charge density. In the following discussions, (which closely follow the papers by *Laughlin* [3.2] and *Morf* and *Halperin* [3.4]) we put the phantom particle at the center of the system for convenience.

For any polynomial $P(z)$, the following identity holds:

$$\left|2\frac{dP}{dz}\right|^2 = \nabla^2|P(z)|^2. \tag{3.17}$$

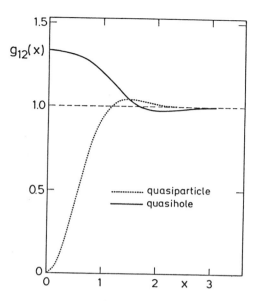

Fig. 3.3. Radial distribution function $g_{12}(x)$ as a function of x for quasiholes and quasiparticles for $m = 3$

The one-particle density is then written

$$\langle \rho(\mathbf{r}_1) \rangle = \frac{\int |\psi|^2 \, d\mathbf{r}_2 \ldots d\mathbf{r}_{N_e}}{\int |\psi|^2 \, d\mathbf{r}_1 \ldots d\mathbf{r}_{N_e}} \quad (3.18)$$

with,

$$|\psi|^2 = \prod_{j=1}^{N_e} \left(e^{mr_j^2} \frac{1}{4m^2} \nabla_j^2 \right) \exp\left[2m \sum_{j<k} \ln |\mathbf{r}_j - \mathbf{r}_k| \right]. \quad (3.19)$$

The denominator of (3.18) is now written as

$$Z \equiv \int |\psi|^2 \, d\mathbf{r}_1 \ldots d\mathbf{r}_{N_e} = \int \prod_{j=1}^{N_e} d\mathbf{r}_j \, e^{-K} \quad (3.20)$$

where,

$$e^{-K} \equiv e^{-\mathcal{H}_m} \prod_{j=1}^{N_e} \left(r_j^2 - \frac{1}{m} \right) \quad (3.21)$$

and $\mathcal{H}_m$ is the one-component plasma Hamiltonian given in (2.27). In (3.20), we have eliminated the Laplacian by performing integration by parts. Similarly,

$$\int |\psi|^2 \, d\mathbf{r}_2 \ldots d\mathbf{r}_{N_e} = \frac{1}{Z} e^{-mr_1^2} \frac{\nabla^2}{4m^2} \int \prod_{j=2}^{N_e} \left[d\mathbf{r}_j \left(r_j^2 - \frac{1}{m} \right) e^{-mr_j^2} \right] \\ \times \exp \left(2m \sum_{j<k} \ln |\mathbf{r}_j - \mathbf{r}_k| \right). \quad (3.22)$$

The final form for the one-particle density is given, after some algebra, by

$$\langle \rho(\mathbf{r}_1) \rangle = \left[\frac{1}{4m^2} \nabla^2 + \frac{1}{m} \mathbf{r} \cdot \nabla_1 + r_1^2 + \frac{1}{m} \right] \frac{\langle \rho^*(\mathbf{r}_1) \rangle}{r_1^2 - \frac{1}{m}} \quad (3.23)$$

with the *integrated by parts* one-particle density $\langle \rho^*(\mathbf{r}_1) \rangle$ given by

$$\langle \rho^*(\mathbf{r}_1) \rangle = \int \prod_{j=2}^{N_e} d\mathbf{r}_j \, e^{-K} \bigg/ \int \prod_{j=1}^{N_e} d\mathbf{r}_j \, e^{-K}. \quad (3.24)$$

This is again a classical plasma problem where the impurity–plasma interaction is given by $v_{12} = -m \ln(r^2 - 2)$ [Eq. (3.21)]. Although the interaction is singular at $r^2 = 2$ and complex for $r^2 < 2$, the HNC method could still be used to calculate $g_{12}(r)$, since the potential enters the equations only in the form, $e^{-\beta v_{12}(r)} = r^2 - 2$. Laughlin also showed that the charge accumulated around the phantom is as expected, $+\frac{e}{m}$.

Since there is no direct plasma analogy in this case, evaluation of the quasiparticle creation energy via HNC is rather subtle. Laughlin introduced the following approximations: suppose there exists a pseudopotential $v_{12}^{ps}(r)$ which gives exactly the same $\langle \rho(\mathbf{r}_1) \rangle$ as given above. He then writes, $g_{12}(r) = \langle \rho(r) \rangle / \rho$ and uses (3.11–14) to evaluate the quasiparticle creation energy. In this manner, he obtained, $\widetilde{\varepsilon}_p = 0.025 \, e^2/\epsilon l_0$ for $m = 3$ and $0.006 \, e^2/\epsilon l_0$ for $m = 5$.

This additional approximation is however *not necessary* if one uses (3.8–10) instead [3.7]. To evaluate $\widetilde{\varepsilon}_p$, we need only $C_{12}^s(x)$ and $\widetilde{u}_{12}^1(q)$ in (3.10). They can be obtained as follows: we write, $u_{12}^s(x) = 2K_0(Qx)$ and invert the HNC-impurity equation (3.7), to obtain,

$$C_{12}^s(x) = [g_{12}(x) - 1] - \ln g_{12}(x) - 2K_0(Qx) \\ N_{12}^s(x) = \ln g_{12}(x) + 2K_0(Qx) \\ \widetilde{u}_{12}^1(q) = \widetilde{C}_{12}^s(q) \widetilde{C}_{11}(q) - \widetilde{N}_{12}^s(q)[1 - \widetilde{C}_{11}(q)]. \quad (3.25)$$

In this manner, the total interaction, $u_{12}(x) = 2K_0(Qx) + u_{12}^1(x)$ would correspond to the $g_{12}(x)$ of (3.10) within the HNC approximation, and no pseudopotential is required. The quasiparticle creation energy was obtained using this method by *Chakraborty* [3.7] with the result, $\widetilde{\varepsilon}_p = 0.025 \, e^2/\epsilon l_0$ for $m = 3$ and $0.0057 \, e^2/\epsilon l_0$ for $m = 5$.

In Fig. 3.4, we present $\widetilde{g}_{12}(x)$ [obtained from $\rho^*(x)$ in (3.24)] evaluated via the two-component classical plasma system for $m = 3$ and $m = 5$. These

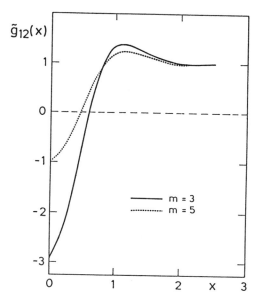

Fig. 3.4. The *integrated by parts* $\widetilde{g}_{12}(x)$ as a function of x for $m = 3$ and $m = 5$

results are similar to those obtained by *Fertig* and *Halperin* [3.5]. The functions $\delta g_{11}(x)$ and $g_{12}(x)$ are plotted for the quasiparticles in Figs. 3.1 and 3.3 respectively. As we shall see below, accurate Monte Carlo results based on Laughlin's quasihole wave functions compare quite well with the quasihole energies obtained above. The quasiparticle creation energy obtained via HNC theory, is however, a factor of two *smaller* than the Monte Carlo result. This point will be discussed in detail in Sect. 3.5.

The quasiparticle and quasihole size is the distance over which the one-component plasma screens [3.1]. For a weakly coupled plasma ($\Gamma \leq 2$), this distance is the Debye length [3.9], $\lambda_D = \frac{\ell_0}{\sqrt{2}}$. For the strongly coupled plasma relevant in the present case, a better estimate is the ion-disk radius associated with *charge* $\frac{1}{m}$, $R = \sqrt{2}\ell_0$. In this sense, the quasiparticles have the same *size* as electrons in the lowest Landau level.

Recently, *Bychkov* and *Rashba* [3.10] have presented an interesting geometrical interpretation of the quasihole and quasiparticle creation operators. In the symmetric gauge, the magnetic translation operators are defined as (see Sect. 4.2)

$$T(a) = t(a) \exp\left[\tfrac{1}{2}i(\widehat{z} \times a) \cdot r\right], \qquad (3.26)$$

where $t(a)$ is the ordinary translation operator and $\widehat{z}$ the unit vector parallel to the magnetic field B. We have chosen units in which $\ell_0 = 1$.

For an *infinitesimal* translation, it is possible to write,

$$T(a) \approx 1 + a \cdot t \tag{3.27}$$

where,

$$t = \nabla + \tfrac{1}{2}i(\hat{z} \times r). \tag{3.28}$$

The complex operators for infinitesimal translations are then,

$$t_\pm = t_x \pm it_y = (\partial_x \pm i\partial_y) \mp \tfrac{1}{2}(x \pm iy). \tag{3.29}$$

Applying these operators to the ground state wave function ψ_m (2.24), we obtain,

$$\begin{aligned} t_+\psi_m &= e^{-|z|^2/4} z^{m+1} \\ t_-\psi_m &= e^{-|z|^2/4}\left(2\frac{d}{dz}z^m\right), \end{aligned} \tag{3.30}$$

which, when compared to (3.2) and (3.16), show that the quasihole and quasiparticle operators are in fact generators of infinitesimal magnetic translations.

Fertig and *Halperin* [3.5] pointed out that for small interparticle separation, $\delta h_{11}(r)$ does not have the correct behavior. It is approximately described by $\delta h_{11} \sim r^{2m}$, whereas the correct behavior should be $\sim r^{2m-4}$. In the presence of an infinitesimal density ρ_2 of quasiparticles at random positions, the electron pair correlation function has the form $g_{11}(|r_1 - r_2|) + \rho_2\, \delta h_{11}(|r_1 - r_2|)$, where $g_{11}(r)$ is the ground state correlation function. Defining the pair correlation function of the electrons in the presence of a quasiparticle at r_0 as

$$g_{112}(r_1 - r_0, r_2 - r_0) = \frac{N(N-1)}{\rho^2} \frac{\int |\psi|^2\, dr_3 \ldots dr_{N_e}}{\int |\psi|^2\, dr_1 \ldots dr_{N_e}} \tag{3.31}$$

the function δh_{11} is written

$$\delta h_{11}(|r_1 - r_2|) = \int dr_0 \left[g_{112}(r_1 - r_0, r_2 - r_0) - g_{11}(|r_1 - r_2|)\right]. \tag{3.32}$$

In the limit $|r_1 - r_2| \to 0$ we have $g_{112}(r_1, r_2) \sim |r_1 - r_2|^{2m-4}$. Therefore, $\delta h_{11}(r) \sim r^{2m-4}$. The g_{112} was then obtained by Fertig and Halperin from an inhomogeneous classical plasma model and it was found that, near a quasiparticle, the electron pair correlation function vanishes as $|r_1 - r_2|^2$ for $|r_1 - r_2| \to 0$. However, their quasiparticle creation energy was a factor of two *larger* than the Monte Carlo result. Accurate evaluation of the correct quasiparticle energy in the plasma approach is still an open problem.

3.2 Finite-Size Studies: Rectangular Geometry

In Sect. 2.1, we have seen that the ground state energy per electron in a finite-size calculation [3.12] shows a cusp-like behavior at $\nu = \frac{1}{3}$. As has already been discussed above, the appearence of a cusp means a positive discontinuity in chemical potential. The chemical potential is defined as

$$\mu = E_0(\nu) + \nu \frac{\partial E_0(\nu)}{\partial \nu} \quad (3.33)$$

where E_0 is the energy per particle at $\nu = \frac{1}{3}$. Adding or subtracting a flux quanta at this filling fraction, we obtain, $\nu_\pm = N_e/(N_s \mp 1)$. Let us define $E_\pm$ to be the ground state energy per particle at those filling fractions. If

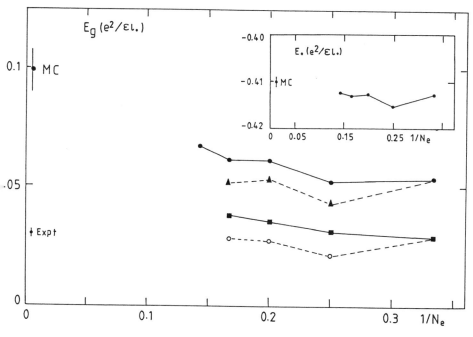

Fig. 3.5. The energy gap for different spin polarization of the quasiparticle (q.p.) and quasihole (q.h.) excitations for three to seven electron systems, in the absence of Zeeman energy. The Monte Carlo result (MC) is from [3.4] for the spin-aligned q.p. + q.h. (●). The other three cases are: spin-aligned q.p. + spin-reversed q.h. (▲); spin-reversed q.p. + spin-aligned q.h. (■); and spin-reversed q.p. + q.h. (O). The experimental data are from [3.11]. The ground state energy values at $\nu = \frac{1}{3}$ [3.4,12] are given in the inset for comparison

we write,

$$\mu_\pm \simeq E_0(\nu) + \nu\left[E_\pm - E_0\right]/(\nu_\pm - \nu) \qquad (3.34)$$

the quasiparticle-quasihole energy gap is then just the difference between the two chemical potentials,

$$E_g = \tfrac{1}{3}(\mu_+ - \mu_-) \qquad (3.35)$$
$$\simeq -2N_e E_0(\nu) + (N_e + \nu) E_0\left(\frac{N_e}{N_s + 1}\right) + (N_e - \nu) E_0\left(\frac{N_e}{N_s - 1}\right).$$

The factor $\tfrac{1}{3}$ in (3.35) is introduced because of the fractional electron charge of the quasiparticles and quasiholes. Using this approach for a four-electron system (spin polarized), *Yoshioka* [3.13] obtained the gap, $E_g \simeq 0.052\, e^2/\epsilon\ell_0$. However, studying somewhat larger systems *Chakraborty* et al. [3.14,15] found that E_g is, in fact, size dependent (see Fig. 3.5) and extrapolation of the results for spin-polarized three- to seven-electron systems (plotted as solid circles) approximately leads to $E_g \simeq 0.1\, e^2/\epsilon\ell_0$. The energy gap can also be estimated from the collective excitation spectrum to be discussed in Chap. 4, and provides essentially the same result.

3.3 Spin-Reversed Quasiparticles

The finite-size calculations discussed above can be readily generalized to the case where the spin of one of the electrons is *reversed* relative to all the others. The spin reversal in the ground state has already been discussed in Sect. 2.5. The same technique will follow through and, in the present case, we will concern ourselves with the two cases, (a) $S = S_z = \tfrac{1}{2}N_e$, which corresponds to the fully spin polarized ground state, and (b) $S = S_z = \tfrac{1}{2}N_e - 1$, which is the spin-reversed excitation we are considering below. Here S and the S_z correspond to the total spin and the z-component of the of the total spin respectively. The spin-reversed quasiparticle and quasihole excitation energy gaps are obtained by evaluating $E_\pm$ in a system where one of the electrons has reversed spin. The ground state energy at $\nu = \tfrac{1}{3}$ [denoted by $E_0(\nu)$ in (3.33–35)] is however calculated for the spin-polarized case.

The major bottleneck of a numerical diagonalization scheme, such as the one described in this section and in the preceding section, is the size of the Hamiltonian matrix; this grows very rapidly with the electron number and at a certain point exceeds the storage capacity of a computer. In the case of one reversed spin, the situation is even worse since the electrons can now occupy states with the same momentum. The matrix dimension in this case is increased approximately by a factor $\sim (1+N_e)$ when compared with that for the fully spin-polarized state. For $N_e > 4$, the dimension of the Hamiltonian matrix is more than 11000, and clearly a straightforward diagonalization is not possible. Furthermore, for the one-spin-reversed problem we are dealing with, the number of electrons and flux quanta is such that they have no common divisor greater than unity. Therefore, the symmetries based on the magnetic translation group (Sect. 4.2) cannot be employed to reduce the matrix size.

Fortunately, there are some simple ways to work with these huge matrices [3.15]. Firstly, the two-body operator of the Hamiltonian (2.14) can connect only those states which differ by at most two indices labelling the occupied single-particle states. The majority of the matrix elements are therefore zero. Moreover, the coefficients of the two-body operator depend only on the difference between the indices and hence there are only a few ($\sim$ number of flux quanta squared) different matrix elements. When the matrix is stored in the computer by rows keeping only non-zero elements, and these are represented as offsets to the array containing the different elements together with the corresponding column indices, only four bytes of storage per non-zero element is required. Since the matrix is symmetric, only the upper or lower triangle need be stored.

The lowest eigenvalue and the corresponding eigenvector are obtained by minimizing the Rayleigh quotient

$$\lambda(x) = \frac{x^T \mathcal{H} x}{x^T x},$$

where x represents the column vector of the coefficients in the superposition of the basis states. Similarly, the next lowest eigenvalue can then be found by working in the subspace orthogonal to this eigenvector. The required number of eigenvalues and eigenvectors can thus be extracted by repeating the procedure. The minimization of the Rayleigh quotient is done by the conjugate gradient method in which the quotient is approximated by a quadratic function and the minimum in each iteration step is searched for in the plane spanned by the gradient and the search direction of the previous iteration step [3.15].

In Fig. 3.5, we have plotted the energy gap for different spin polarizations of the quasiparticles and quasiholes. The spin-polarized quasiparticle–quasihole case has been discussed earlier in Sect. 3.2. The other three cases are: (a) spin-polarized quasiparticle and spin-reversed quasihole gap (plotted as ▲), (b) spin-reversed quasiparticle and spin-polarized quasihole (plotted as ■) and (c) spin-reversed quasiparticle and quasihole (plotted as ○). Obviously, the lowest energy excitations *in the absence of Zeeman energy*, at $\nu = \frac{1}{3}$ involve spin reversal. The origin of this could perhaps be traced

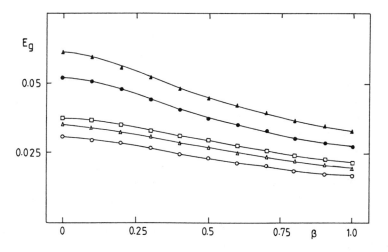

Fig. 3.6. Energy gap E_g as a function of the dimensionless parameter $\beta = (b\ell_0)^{-1}$ for four (● and ○), five (▲ and △) and six (□) electron systems. The filled circles are for spin-polarized q.p. + q.h. case, while the open circles are for the spin-reversed q.p. + spin polarized q.h. case. The five and six electron results in the former case are not distingushable in the figure

to the results in Sect. 2.5. We found that for any state other than $\frac{1}{m}$ with m an odd integer (which is spin polarized, in agreement with the Laughlin state), the electron-electron interaction decreases the tendency to spin polarization, resulting in lower energies for the spin reversed cases considered here. The spin reversal is found to cost less energy for E_+ (reduction of a flux quanta) than that for E_- (addition of a flux quanta). Therefore, in the absence of Zeeman energy, the lowest energy excitations correspond to the case where E_+ and E_- are evaluated for spin-reversed systems.

In Fig. 3.6, we present the energy gap E_g as a function of the dimensionless parameter $\beta = (b\ell_0)^{-1}$ (see Sect. 2.6) for four- and six-electron systems at $\nu = \frac{1}{3}$. For the spin-polarized quasiparticle–quasihole case (filled points), five- and six-electron system results are indistingushable in the figure. The interesting point, however, is the fact that there is a substantial reduction of the gap in the range $\beta = 0.5$–1.0, compared to the ideal case of $\beta = 0$. In fact, the ratio $E_g(\beta = 1)/E_g(\beta = 0)$ is 0.53 and 0.52 for the four- (●) and five-electron (▲) systems respectively. In the case of spin-reversed quasiparticle and spin-polarized quasihole (empty points), the reduction is slightly less: 0.55, 0.56 and 0.57 for four- (○), five- (△) and six-electron (□) systems respectively [3.14,15]. Similar reduction of the energy gap was also observed by *Zhang* and *Das Sarma* [3.16] for finite-size systems. The cases (a) and (c) are not shown in Fig. 3.6, since, as we shall see in Sect. 3.6, they are not energetically favored when the Zeeman energy is added to these results and when compared with the experimental results. In fact, it will be seen in Sect. 3.6 that for small magnetic fields, the spin-reversed quasiparticles are clearly favored energetically.

For an infinite system, *Morf* and *Halperin* [3.4] proposed the following trial wave function for spin-reversed quasiparticles

$$\psi = \prod_{j=2}^{N_e} (z_j - z_1)^{-1} \psi_m. \qquad (3.36)$$

Bringing in the plasma analogy by writing the probability density as

$$|\psi|^2 = e^{-\mathcal{H}},$$

the plasma Hamiltonian can now be written as

$$\mathcal{H} = -2m\left(1 - \tfrac{1}{m}\right) \sum_{j=2}^{N_e} \ln|z_j - z_1| - 2m \sum_{\substack{j<k \\ j>1}}^{N_e} \ln|z_j - z_k| + \sum_{j=1}^{N_e} |z_j|^2 / 2\ell_0^2. \qquad (3.37)$$

In this case, $|\psi|^2$ is the distribution function for a two-dimensional plasma in which particle 1 has its charge *reduced* by a factor $(1 - \tfrac{1}{m})$ in its repul-

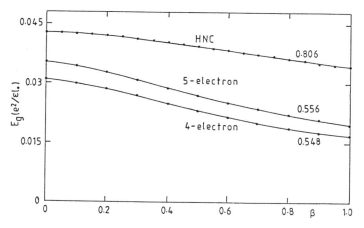

Fig. 3.7. The energy gap E_g (in units of $e^2/\epsilon\ell_0$) for the case of spin-polarized quasihole and spin-reversed quasiparticle as a function of β in the HNC approximation. The four- and five-electron system results are from [3.14]. The number attached to each curve indicates the ratio of $\beta = 1$ and $\beta = 0$ values of E_g

sive interaction with the other particles. The particle 1 has, however, the same interaction as the other particles in its attractive interaction with the background. Particle 1 will therefore be attracted to the center of the disk, while a two-dimensional *bubble* will be formed near the origin of size $(1-\frac{1}{m})$. As a result, there will be an extra negative charge $\frac{e}{m}$ near the origin. Furthermore, when (3.36) is considered to be a function of the position of any electron other than the singled-out electron 1, there will be one less zero of the wave function than in the case of the ground state ψ_m.

The HNC calculation for the wave function defined above was performed by *Chakraborty* [3.8]. In Fig. 3.7, we present the results for the energy gap for the spin-reversed quasiparticle and spin-polarized quasihole and compare these with the corresponding cases for finite-size systems discussed above. Both types of calculation provide qualitatively similar results.

3.4 Spherical Geometry

In Sect. 2.3, we discussed the ground state calculations by *Haldane* and *Rezayi* [3.17] in a spherical geometry. These authors also computed the energy gap in three- to seven-electron systems at $\nu = \frac{1}{3}$. Later, *Fano* et al. [3.18] performed similar calculations for larger systems at $\nu = \frac{1}{3}$ and $\frac{1}{5}$. The quasiparticle and quasihole wave functions on the sphere are written

$$\psi_{\text{p}} = \prod_i \frac{\partial}{\partial v_i} \psi_m \qquad (3.38)$$

$$\psi_{\text{h}} = \prod_i v_i \psi_m, \qquad (3.39)$$

respectively, where ψ_m is the ground state wave function ($L = 0$). These states have total angular momentum $L = \frac{1}{2} N_{\text{e}}$, and azimuthal angular momentum $M = \pm L$.

The numerical results by Haldane and Rezayi for the quasiparticle–quasihole gap for three- to seven-electron systems are shown in Fig. 3.8. Their estimate for the infinite system is $E_{\text{g}} \approx 0.105 \pm 0.005\, e^2/\epsilon \ell_0$. The charge density profiles of the quasiparticle and quasihole obtained for a six-electron system is shown in Fig. 3.9. As discussed earlier, creation of quasiparticles and quasiholes at fixed total charge would imply that the charge of the background condensate must be decreased or increased by $\frac{e}{m}$, because of the fractional electron charge of the quasiparticles and quasiholes

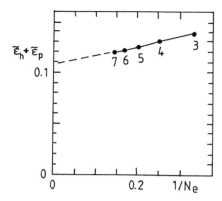

Fig. 3.8. The energy required to create a quasihole plus quasiparticle pair at infinite separation, evaluated in the spherical geometry [3.17]

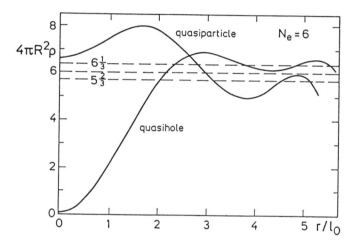

Fig. 3.9. Charge density profiles of a quasiparticle and quasihole [3.17]

(see Sect. 3.1). The asymptotic charge densities in Fig. 3.9 seem to approach the value $4\pi R^2 \rho \approx N \mp \frac{1}{3}$.

Fano et al. [3.18] have performed a similar calculation for somewhat larger systems and claimed to have improved on the results of Haldane and Rezayi. In Table 3.1, we present their results for the quasihole and quasiparticle creation energies and the energy gap for $\nu = \frac{1}{3}$ and $\frac{1}{5}$.

55

Table 3.1. Quasiparticle and quasihole creation energies and the energy gap for finite electron systems in a spherical geometry. HR represents results of [3.17]. The other results are from [3.18]. The unit of energy is $e^2/\epsilon\ell_0$.

N_e	$\nu = \frac{1}{3}$				$\nu = \frac{1}{5}$		
	$\tilde{\varepsilon}_h$	$\tilde{\varepsilon}_p$	E_g	E_g(HR)	$\tilde{\varepsilon}_h$	$\tilde{\varepsilon}_p$	E_g
3	0.04270	0.11968	0.16238	0.13849	0.01269	0.02787	0.04057
4	0.03782	0.10469	0.14251	0.13093	0.01055	0.02337	0.03392
5	0.03549	0.09896	0.13445	0.12543	0.01034	0.02257	0.03290
6	0.03362	0.09363	0.12726	0.12045	0.00849	0.01970	0.02819
7	0.03257	0.09101	0.12359	0.11816	0.00907	0.02019	0.02925
8	0.03172	0.08877	0.12049	—	—	—	—
9	0.03103	0.08700	0.11803	—	—	—	—
∞	0.02640	0.07720	0.10360	0.105±0.005	0.00710	0.01730	0.02440

3.5 Monte Carlo Results

For the quasihole and quasiparticle excitation energies, extensive Monte Carlo calculations have been performed by *Morf* and *Halperin* [3.4,6,19]. Let us begin with the results for the quasihole. The results for the electron density $\langle \rho(r) \rangle$ generated via the Monte Carlo simulation of the plasma Hamiltonian (3.4) with a quasihole at the origin ($z_0 = 0$) are shown in Fig. 3.10 for $\nu = \frac{1}{3}$. This density, and the ones that will follow correspond

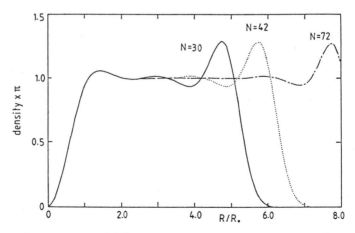

Fig. 3.10. Electron density $\langle \rho(r) \rangle$ for Laughlin's quasihole state at $\nu = \frac{1}{3}$ for N_e=30, 42 and 72 electrons [3.4]

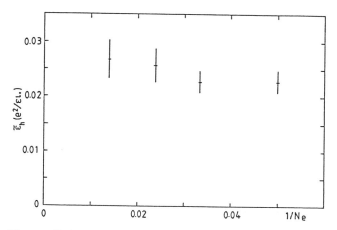

Fig. 3.11. The quasihole creation energy, $\tilde{\varepsilon}_h$ for $\nu = \frac{1}{3}$ with Laughlin's quasihole wave function [3.4]

to a system where the change in the occupied area due to the presence of a quasihole or a quasiparticle is not compensated. For small interparticle separation, the results are essentially independent of particle number and are similar to those of Fig. 3.3 for the HNC approximation. The particle excess,

$$X(r) = 2\pi \int [\rho(r) - \rho_m] r \, dr, \qquad (3.40)$$

where $\rho_m = (2\pi \ell_0^2 m)^{-1}$, is evaluated to be, $-0.34, -0.326$, and -0.327 at $R = 2.5$ for the 30-, 42-, and 72-particle systems respectively.

The quasihole creation energy is computed for 20-, 30-, 42-, and 72-electron systems and is plotted in Fig. 3.11. The standard deviation is indicated by the length of the vertical bars in the figure. Due to large statistical errors, reliable extrapolation to the thermodynamic limit is difficult. The result for the 72-electron system, $\tilde{\varepsilon}_h = 0.0268 \pm 0.0033 \, e^2/\epsilon \ell_0$ is already close to the HNC result.

For Laughlin's quasiparticle wave function, we have mentioned earlier that a direct plasma analogy is not possible. A plasma analogy exists however for the *integrated by parts* one-particle density $\langle \rho^*(\mathbf{r}_1) \rangle$ [see (3.24)] for the Hamiltonian K in (3.21). In Fig. 3.12, we show a plot of $\rho^*(\mathbf{r}) = (r^2 - \frac{1}{m})\tilde{f}(r)$, where the solid line is an interpolation of the Monte Carlo results for 30 electrons (dots), using a Padé approximation,

$$\tilde{f}(r) \simeq (a_0 + a_1 r^2 + a_2 r^4)/(1 + b_1 r^2),$$

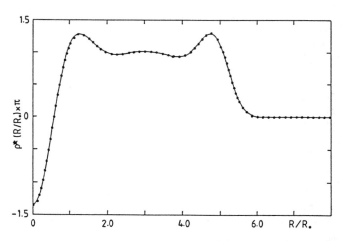

Fig. 3.12. Unphysical density $\rho^*(r)$ for the Hamiltonian K [see (3.21)] for Laughlin's quasiparticle state at $\nu = \frac{1}{3}$ [3.4]

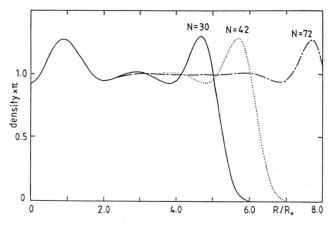

Fig. 3.13. Electron density $\langle \rho(r) \rangle$ for Laughlin's quasiparticle state at $\nu = \frac{1}{3}$ [3.4]

and a third-order spline polynomial for $r > 1.2$. The Padé coefficients are $a_0 = 4.05112, a_1 = -2.08448, a_2 = 0.38863$ and $b_1 = 0.42435$. Using this fit for $\widetilde{f}(r)$ in (3.23), *Morf* and *Halperin* [3.4] obtained the final result for the density $\langle \rho(r) \rangle$ as shown in Fig. 3.13.

It is obvious in Fig. 3.13 that the excess charge is located inside a circle of radius $R \approx 2$. There is a pronounced dip at the origin, which is not seen in the plasma approach (Fig. 3.3). A dip is also present in the quasiparticle charge density evaluated on the sphere (Fig. 3.9). For the 72-particle system,

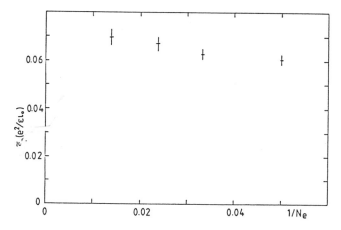

Fig. 3.14. Quasiparticle creation energy $\widetilde{\varepsilon}_{\mathrm{p}}$ for the Laughlin wave function at $\nu = \frac{1}{3}$ [3.4]

the particle excess is evaluated as 0.207, 0.408, and 0.33 at $R=1$, 2, and 3 respectively. For $3 < R < 6$, $X(R)$ oscillates around $\frac{1}{3}$, as required.

The quasiparticle creation energy obtained for 20, 30, 42, and 72 particle systems with Laughlin's wave function are shown in Fig. 3.14. For the 72-electron system, *Morf* and *Halperin* [3.4] obtained the result $\widetilde{\varepsilon}_{\mathrm{p}} = 0.0698 \pm 0.0033\, e^2/\epsilon\ell_0$. Reliable extrapolation of the results in Fig. 3.14 to the thermodynamic limit is difficult. However these authors made a plausible estimate of $\widetilde{\varepsilon}_{\mathrm{p}} \approx 0.073 \pm 0.008\, e^2/\epsilon\ell_0$. The result is significantly higher than the estimate from the plasma calculation (see Sect. 3.1). The energy gap to create a quasiparticle–quasihole pair far apart from each other is then $E_{\mathrm{g}} \approx 0.099 \pm 0.009\, e^2/\epsilon\ell_0$ in the thermodynamic limit.

An alternative trial wave function for the quasiparticles, which is more directly motivated by a classical statistical-mechanics problem, has been proposed by *Halperin* [3.3]. The Monte Carlo evaluation of the quasiparticle energy and of the correlation function using this wave function has been reported by *Morf* and *Halperin* [3.4,19]. Placing the quasiparticle at the origin, the wave function is written

$$\psi_0^{(+)}\{z_k\} = \mathcal{A}\widetilde{\psi}_0^{(+)}\{z_k\}$$

$$\widetilde{\psi}_0^{(+)}\{z_k\} = \left[\frac{1}{z_1 - z_2}\right]^2 \prod_{j=3}^{N_e} \frac{z_j - \frac{1}{2}(z_1 + z_2)}{(z_1 - z_j)(z_2 - z_j)} \psi_m \quad , \tag{3.41}$$

$\mathcal{A}$ being the antisymmetrization operator. It is readily noted that, $\widetilde{\psi}_0^{(+)}$ has the form of a polynomial in $\{z_k\}$, multiplied by a gaussian factor, and therefore describes a collection of particles in the lowest Landau level, for $m > 1$. The plasma analogy in this case results in the following Hamiltonian:

$$|\widetilde{\psi}_0^{(+)}|^2 = e^{-\widetilde{\mathcal{H}}}$$

$$\widetilde{\mathcal{H}} = \mathcal{H}_m + 4\ln|\mathbf{r}_1 - \mathbf{r}_2| + 2\sum_{j=1}^{N_e}\Big(\ln|\mathbf{r}_j - \mathbf{r}_1| + \ln|\mathbf{r}_j - \mathbf{r}_2| \\ - \ln|\mathbf{r}_j - \tfrac{1}{2}(\mathbf{r}_1 + \mathbf{r}_2)|\Big). \tag{3.42}$$

where $\mathcal{H}_m$ is the one-component plasma Hamiltonian (2.27). We thus have an extra logarithmic attraction between particles 1 and 2, resulting in a bound pair. The other particles experience a logarigthmic repulsive interaction with the center of gravity of the bound pair, and a logarithmic attraction to the two members of the pair. As a result, an electron somewhat away from the pair, would see a net charge of $(2m - 1)$ on the pair, in units where an unpaired electron has charge m. Therefore, we have a hole of size $(2 - \frac{1}{m})$ about the pair. Including the pair, there will be a net charge of $\frac{e}{m}$.

The wave function $\widetilde{\psi}_0^{(+)}$ is antisymmetric with respect to interchange of particles j and k with $j > 2$ and $k > 2$. It is also antisymmetric with respect to particles 1 and 2. It does not, however, change sign under a permutation, for example P_{13}, that interchanges positions of particles 1 and 3. If the pair of particles 1 and 2 were very tightly bound relative to the separation between pairs, and had no overlap in space with the region occupied by particle 3, $P_{13}\widetilde{\psi}_0^{(+)}$ would have no overlap with $\widetilde{\psi}_0^{(+)}$. There would be no contribution of $\langle\widetilde{\psi}_0^{(+)}|P_{13}\widetilde{\psi}_0^{(+)}\rangle$ to the normalization of $\widetilde{\psi}_0^{(+)}$ and no contribution from $\langle\widetilde{\psi}_0^{(+)}|V|P_{13}\widetilde{\psi}_0^{(+)}\rangle$ to the expectation value of the potential energy. As, $P_{13}^2 = 1$, and P_{13} commutes with V, it is clearly seen that the following relations hold,

$$\langle\widetilde{\psi}_0^{(+)}|P_{13}VP_{13}|\widetilde{\psi}_0^{(+)}\rangle = \langle\widetilde{\psi}_0^{(+)}|V|\widetilde{\psi}_0^{(+)}\rangle$$
$$\langle\widetilde{\psi}_0^{(+)}|P_{13}P_{13}|\widetilde{\psi}_0^{(+)}\rangle = \langle\widetilde{\psi}_0^{(+)}|\widetilde{\psi}_0^{(+)}\rangle$$

and hence the expectation value of the potential energy would be unaffected. In actual practice however, we do not expect a zero overlap of the pair (z_1, z_2) with other other electrons in the system. We can hope that the antisymmetrizer $\mathcal{A}$ has only a modest effect on the energy and correlation functions for the quasiparticle state.

The mean density, $\langle\widetilde{\psi}_0^{(+)}|\rho(\mathbf{r})|\widetilde{\psi}_0^{(+)}\rangle$ obtained by Monte Carlo simulation of a system described by the Hamiltonian $\widetilde{\mathcal{H}}$ (3.42) is shown in Fig. 3.15

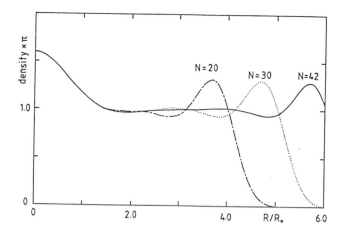

Fig. 3.15. The mean density $\langle \rho(r) \rangle$ in the state with quasiparticle excitation at $\nu = \frac{1}{3}$. The Monte Carlo result is obtained for the *non*antisymmetrized pair wave function $\widetilde{\psi}_0^{(+)}$ [3.4]

for 20, 30, and 42 particles. The quasiparticle energy for a fully antisymmetrized pair wave function $\psi_0^{(+)}$ obtained for 42 electrons is $0.066 \pm 0.006\, e^2/\epsilon\ell_0$. This result is almost identical (within the statistical uncertainty) to the 42-particle system result from Laughlin's wave function.

Monte Carlo computations of the quantities described above, were also performed for the spherical geometry by *Morf* and *Halperin* [3.19]. For the quasihole state (3.39), energies were evaluated for the two different definitions of distance. For the chord distance, they obtained

$$\widetilde{\varepsilon}_h \approx 0.0219(25) - 0.055(30)/N_e \tag{3.43}$$

and for the great circle distance

$$\widetilde{\varepsilon}_h \approx 0.0229(21) - 0.078(24)/N_e. \tag{3.44}$$

For the bulk limit, their best estimate is, $\widetilde{\varepsilon}_h \approx 0.0224 \pm 0.0016\, e^2/\epsilon\ell_0$.

In the case of the quasiparticle state (3.38), methods similar to those described in Sect. 3.1 were used. The quasiparticle energy $\widetilde{\varepsilon}_p$ at $\nu = \frac{1}{3}$ is displayed in Fig. 3.16 for various values of system size. The *pair* wave function discussed above for disk geometry, is written for the spherical geometry as

$$\psi^{(+)} = \mathcal{A}\widetilde{\psi}^{(+)}, \tag{3.45}$$

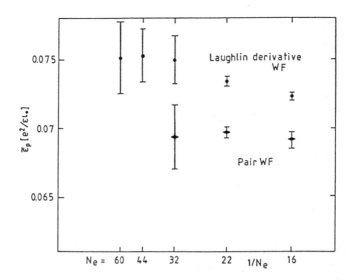

Fig. 3.16. Quasiparticle energy $\widetilde{\varepsilon}_p$ at $\nu = \frac{1}{3}$ for the Laughlin wave function (full circles) and the fully antisymmetrized pair wave function (diamonds) as a function of $\frac{1}{N_e}$ [3.19]

with

$$\widetilde{\psi}^{(+)} = \frac{u_1 u_2}{(u_1 v_2 - v_1 u_2)^2} \psi_m \prod_{j=3}^{N_e} \frac{2v_j u_1 u_2 + u_j(u_1 v_2 + v_1 u_2)}{(u_1 v_j - v_1 u_j)(u_2 v_j - v_2 u_j)}, \quad (3.46)$$

where u_j, v_j are, as usual, the spinor coordinates of the jth electron. The quasiparticle energy for the fully antisymmetrized pair wave function is also presented in Fig. 3.16. The energies in this case are 4-6% lower than those for the Laughlin state. In the thermodynamic limit, the best estimate is [3.19] $\widetilde{\varepsilon}_p \approx 0.070 \pm 0.003\, e^2/\epsilon\ell_0$ for the pair wave function, and $\widetilde{\varepsilon}_p \approx 0.075 \pm 0.005\, e^2/\epsilon\ell_0$ for the Laughlin wave function. The energy gap in the spherical geometry at $\nu = \frac{1}{3}$ is $E_g \approx 0.092 \pm 0.004\, e^2/\epsilon\ell_0$. Including the finite thickness correction (see Sect. 2.6) the excitation energies as a function of the dimensionless thickness parameter β are as plotted in Fig. 3.17.

The density in the quasiparticle state is shown in Fig. 3.18 for the two wave functions as a function of the great circle distance from the north pole ($r = R/R_0$, R_0 being the ion disk radius) for a 32-electron system. For the fully antisymmetrized pair wave function (open circles) and for the Laughlin wave function (solid line), there is a strong dip at the origin. For the nonantisymmetrized wave function, $\rho(r)$ has its maximum at the origin (dashed line).

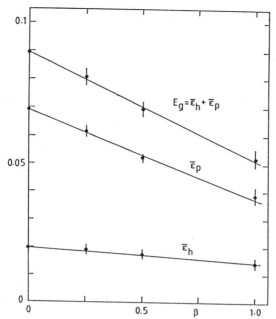

Fig. 3.17. Excitation energies at $\nu = \frac{1}{3}$ versus the dimensionless thickness parameter $\beta = (b\ell_0)^{-1}$. The quasihole results are for the Laughlin wave function, while the quasiparticle results are for the fully antisymmetrized pair wave function. The results in this figure are for a 22 electron system [3.19]. The unit of energy is $e^2/\epsilon\ell_0$

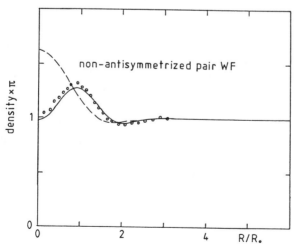

Fig. 3.18. Density $\rho(r)$ in the quasiparticle state at $\nu = \frac{1}{3}$. The solid line corresponds to the Laughlin wave function, while the open circles denote results for the fully antisymmetrized pair wave function. The dashed line is for the pair wave function without the antisymmetrizer. The results are for a 32-electron system [3.19]

MacDonald and *Girvin* [3.20,21] proposed a somewhat different approach to calculate the density and energy for the quasiparticle and quasihole state. Their results for the energies at $\nu = \frac{1}{3}$ are, $\widetilde{\varepsilon}_h = 0.0287\, e^2/\epsilon \ell_0$ for the quasihole and $\widetilde{\varepsilon}_p = 0.085\, e^2/\epsilon \ell_0$ for the quasiparticles. Their quasiparticle density did not show any dip at the origin. These authors also calculated the above quantities in the higher Landau level, which will be discussed in Sect. 6.2.

In the spherical geometry, *Morf* and *Halperin* also calculated the energy and density for the *spin-reversed* quasiparticle state at $\nu = \frac{1}{3}$ [3.19]. In this case, the wave function (3.36) is written as

$$\psi = \psi_m u_1^{-1} \prod_{j=2}^{N_e} \left(\frac{u_1}{u_1 v_j - v_1 u_j} \right). \tag{3.47}$$

The energy gap for a spin-reversed quasiparticle and a spin-polarized quasihole was estimated to be, $E'_g \approx 0.063 \pm 0.005\, e^2/\epsilon \ell_0$, significantly larger than the exact result $E'_g = 0.037\, e^2/\epsilon \ell_0$ for six-electron system by *Chakraborty* et al., [3.14] (see Sect. 3.3). The density $\rho(r)$ in the spin-reversed quasiparticle state at $\nu = \frac{1}{3}$ could be fitted approximately by the following gaussian:

$$\rho(r) \approx \frac{1}{\pi} \left[1 + 0.46\, e^{-(r/0.851)^2} \right]. \tag{3.48}$$

No dip at the origin and no local minimum was found to appear around $r \approx 2$.

To summarize this section, while the calculated quasihole creation energy is very similar for three different approaches (plasma; Monte Carlo, disk; Monte Carlo, sphere), there is a big discrepancy between the classical plasma result and the Monte Carlo results in the case of the quasiparticle energy. The Monte Carlo results for the disk geometry and the spherical geometry are however consistent. The quasiparticle creation energy is larger than the quasihole creation energy by a factor of three. The energy gap for a quasiparticle and quasihole pair at infinite separation appears from all Monte Carlo estimates to be $E_g \approx 0.1\, e^2/\epsilon \ell_0$, which is close to the estimates from finite system calculations. The numerical results for the density also suggest that there is a strong dip at the origin; this is missing in the plasma approach.

3.6 Experimental Investigations of the Energy Gap

After reviewing the theoretical work on the quasiparticle–quasihole gap E_g in the FQHE state, we would like to present in this section, a brief review of the experimental investigations of the energy gap. As mentioned in the introduction, in both the integer and fractional QHE, the vanishing of the diagonal resistivity implies a gap in the excitation spectrum. In the case of the integer QHE, the gap is in the single-particle density of states, whereas in the FQHE, the gap lies in the excitation spectrum of the correlated many-electron ground state. The energy gap is usually obtained from the temperature dependence of the magnetoconductivity, σ_{xx} [or ρ_{xx} since near the ρ_{xx} minima, $\rho_{xx} \ll \rho_{xy}$, and $\sigma_{xx} = \rho_{xx}/(\rho_{xx}^2 + \rho_{xy}^2) \sim \rho_{xx}/\rho_{xy}^2$], as $\sigma_{xx} \propto \rho_{xx} \propto \exp(-W/k_B T)$, where $W = \frac{1}{2}E_g$ is the activation energy, and k_B is Boltzmann's constant. In the case of the integer QHE, *Tausendfreund* and *von Klitzing* found the energy gap obtained from the activation energy measurements to be close to the cyclotron energy (usually smaller because of spin splitting and a finite linewidth of the extended states) [3.22]. For the FQHE, similar measurements for the energy gap have been undertaken by several experimental groups.

The first such work was by *Chang* et al. [3.23] for $\nu = \frac{2}{3}$, whose results indicated that the energy gap does not have a $B^{\frac{1}{2}}$ behavior and also suggested the existence of a mobility threshold below which the $\frac{2}{3}$ effect will not occur. In the theoretical calculations, the energy gap scales with the natural unit of energy $e^2/\epsilon\ell_0$, and hence should have a magnetic field dependence of $B^{\frac{1}{2}}$. This behavior is however changed, as we shall see below, when the finite thickness correction is included in the calculations. A mobility threshold for $\nu = \frac{4}{3}$ was also observed in Si-MOSFETs by *Kukushkin* and *Timofeev* [3.24,25].

A systematic study of the energy gap for the filling fractions, $\nu = \frac{1}{3}, \frac{2}{3}, \frac{4}{3}$, and $\frac{5}{3}$ was reported by *Boebinger* et al. [3.26,27]. The study was performed in four specimens of modulation-doped GaAs-heterostructure with typical mobilities of μ=5 000 000–850 000 cm^2/V.s and electron densities, $\rho_0 = (1.5 - 2.3) \times 10^{11}$cm^{-2}. The value of ρ_{xx} or σ_{xx} at the minimum corresponding to a particular filling fraction was determined as a function of temperature in the range of 120 mK to 1.4 K. Figure 3.19 shows the activated conduction in the case of $\nu = \frac{2}{3}$ for different values of the magnetic field. At high temperatures, ρ_{xx} deviates from a simple activated dependence. For magnetic fields between 6 and $\sim$ 10 T, simple activated behavior was observed. Data for $B \geq 10$ T showed deviation from simple activated behavior even at low temperatures. Most of the high-field data behaved as shown in Fig. 3.19(b), where the deviation is curved and smooth. Over the entire temperature range, activation plus various models for hopping con-

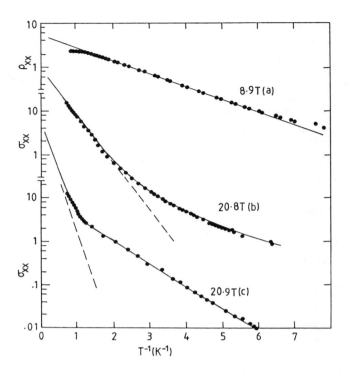

Fig. 3.19. Temperature dependence of $\sigma_{xx}(10^{-7}\Omega^{-1})$ and $\rho_{xx}(10^3\Omega/\square)$ at the minimum for $\nu = \frac{2}{3}$ at (a) B=8.9 T, (b) B=20.8 T, and (c) B=20.9 T [3.26]

duction was found to fit the data very well. One of the samples showed the behavior depicted in Fig. 3.19(c), where the deviation from a simple activation is a sharp break to a second linear region. A similar case was observed by *Kawaji* et al. [3.11,28], who interpreted the results as the existence of two different activation energies. Boebinger et al., however, explained the result as an artifact caused by a nonequilibrium configuration of the electronic state within the sample.

The activation energies for the filling factors $\nu = \frac{1}{3}, \frac{2}{3}, \frac{4}{3}$, and $\frac{5}{3}$ are presented in Fig. 3.20. The following features are noteworthy in the result: (i) No apparent sample dependence was observed. (ii) The data for $\nu = \frac{1}{3}$ and $\frac{2}{3}$ overlap at $B \sim 20$ T. At similar magnetic fields, the data for $\nu = \frac{4}{3}$ and $\frac{5}{3}$ are consistent with the data for $\nu = \frac{2}{3}$ at similar magnetic fields. Collectively, the results therefore suggest a *single* activation energy $^3\Delta$, for all the filling fractions mentioned above. A curve for $^3\Delta = Ce^2/\epsilon\ell_0$ with $C = 0.03$ is plotted in Fig. 3.20 for comparison. (iii) The observed activation energies are much *smaller* than the theoretical predictions discussed in the earlier sections. (iv) As mentioned above, $^3\Delta$ does not follow the expected $B^{\frac{1}{2}}$

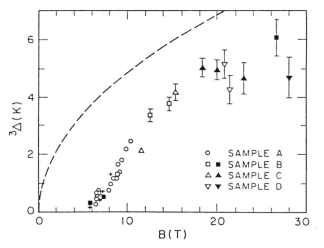

Fig. 3.20. Activation energies for the thirds (in units of K) [3.26,27] as a function of magnetic field. Open symbols are the data for $\nu = \frac{2}{3}$. Filled symbols are for $\nu = \frac{1}{3}$, except for two filled squares at 5.9T and 7.4T, which are for $\nu = \frac{5}{3}$ and $\frac{4}{3}$. The four data points shown as (+) are from [3.29] and [3.30]

dependence. For $B \lesssim 5.5$ T, $^3\Delta$ is vanishingly small. For higher magnetic fields, there is a roughly *linear* increase in $^3\Delta$ up to ~ 12 T.

In Sect. 3.2, we reviewed the results for the energy gap, E_g, from finite-size calculations in a periodic rectangular geometry and noticed that spin-reversal for the quasiparticles cost less energy (in the absence of Zeeman energy) compared to fully spin-polarized quasiparticles and quasiholes. In the following, we would like to compare those results with the observed values of $^3\Delta$. At $\nu = \frac{1}{3}$, the magnetic field dependence of the finite-thickness parameter β (Sect. 2.6) is $\beta = 0.525 B^{\frac{1}{6}}$. The Zeeman energy contribution (per particle) to the energy gap for the cases considered in Sect. 3.2 are (a) zero for the fully spin-polarized quasiparticle–quasihole case, (b) $\left(1 + \frac{1}{3N_e}\right)\varepsilon_z$ for the spin-reversed quasihole and spin-polarized quasiparticle case, (c) $\left(1 - \frac{1}{3N_e}\right)\varepsilon_z$ for the spin-reversed quasiparticle and spin-polarized quasihole case, (d) $2\varepsilon_z$ for the spin-reversed quasiparticle and quasihole case. Here $\varepsilon_z = g\mu_B B$ (Sect. 2.5), with μ_B the Bohr magneton and the Landé g-factor, $g \simeq 0.52$ for GaAs. With B given in Tesla, we obtain $\varepsilon_z = 0.355 B$ in Kelvin.

The magnetic field dependence of the energy gap E_g for a five-electron system is shown in Fig. 3.21, where we have presented only the lowest energy results.[3] For low magnetic fields, the curve for the lowest energy excitations

---

[3] In this figure the experimental results contain some additional data not present in Fig. 3.20 (G. S. Boebinger, private communications, and [3.31]).

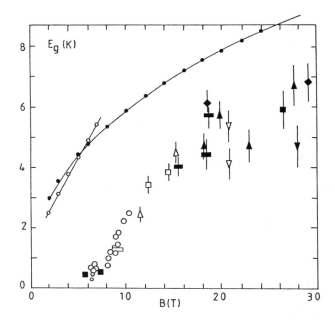

Fig. 3.21. Energy gap E_g (in units of K) vs magnetic field B (given in Tesla) for a five-electron system. The empty points in the theoretical curves are for spin reversed quasiparticles and spin polarized quasiholes, while the filled points are for the fully spin polarized case

rises *linearly* as a result of the *spin-reversed quasiparticles* [case (c) above], which include the dominant contribution from the Zeeman energy, itself linear in magnetic field. As the magnetic field is increased, a crossover point is reached, beyond which the $B^{\frac{1}{2}}$ dependence (modified by the magnetic field dependence of B) is then obtained due to the spin-polarized quasiparticles and quasiholes [3.14,15]. While these theoretical results are open to improvement, the point should be made that, for low magnetic fields, the spin-reversed quasiparticles do play an important role in the elementary excitations in the FQHE. The observation of a threshold field of ~ 5.5 T is not explained, however, by these theories.

A systematic study of the influence of disorder on the activation gap has been made by *Kukushkin* and *Timofeev* [3.24,25]. Their study was based on Si-MOSFETs with high mobility of the two-dimensional electron system in the inversion layer. In this case, it is possible to determine the activation energy within the same structure for different filling factors, while the other parameters (e.g. the electron mobility μ_e) and the magnetic field are fixed. For different MOS structures, they found that the activation gaps for $\nu = \frac{1}{3}, \frac{2}{3}, \frac{4}{3}$ and $\frac{4}{5}$ increase as $B^{\frac{1}{2}}$ in the region B=10–20 T (see Fig. 3.22).

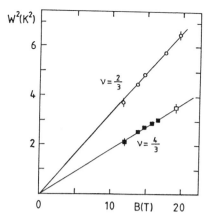

Fig. 3.22. Activation gap W as a function of magnetic field at $\nu = \frac{2}{3}$ and $\frac{4}{3}$ measured for two Si-MOSFETs (closed and open points) with mobility $\mu_e = (3.5 \pm 0.1) \times 10^4 \text{cm}^2/\text{V.s}$ (circles) and $\mu_e = (2.7 \pm 0.1) \times 10^4 \text{cm}^2/\text{V.s}$ (squares) [3.25]

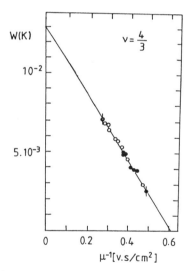

Fig. 3.23. The activation energy as a function of the reciprocal electron mobility at $\nu = \frac{4}{3}$. The filled and empty points are for two different MOSFETs [3.25]

The magnitude of activation gaps, as mentioned earlier, also depend on the mobility. For a given magnetic field, the gaps increase with mobility. The gap W, as a function of mobility can be expressed as (see Fig. 3.23),

$$W_\nu = G_\nu^\infty \left(1 - \frac{\mu_0}{\mu_e}\right) \frac{e^2}{\epsilon \ell_0}$$

69

where, G_ν^∞ is the activation gap when $\mu \to \infty$, and μ_0 is the minimal mobility when the activation energy vanishes (for $W_\nu \to 0$, FQHE is not observed). Experimentally, it is found that μ_0 does not depend on the magnetic field.

The effect of disorder on the activation gap has been studied theoretically by *MacDonald* et al. [3.32]. In particular, they examined the contribution of remote ionized impurities on the disorder potential. The influence of impurities on the conductivity was treated in the lowest order perturbation theory via a memory function approach. Associating the experimentally observed activation energy with the minimum energy of the disordered band of magnetoroton excitations, they obtained a fit to the experimental results of [3.26], with a qualitatively correct behavior for the disorder threshold. It is not clear however, how the magnetoroton[4] — an electrically neutral object—would explain the magnetotransport measurements. The effect of

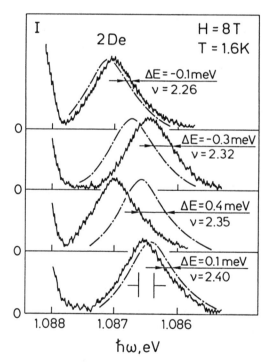

Fig. 3.24. Recombination spectra for two-dimensional electrons measured at T=1.6 K and T=4 K (dashed line) for different values of ν: 2.26, 2.32, 2.35, 2.40 (B=8 T). Here ΔE is the difference in the spectral position of the lines, measured at T=1.6 K and T=4 K [3.36]

---

[4] A detailed discussion of magnetorotons is given in Sect. 4.4.

disorder on the activation energy gap has also been studied by *Gold* [3.33]. Both of these theoretical approaches, however, contain adjustable parameters in order to fit the experimental results by *Boebinger* et al. [3.26,27]. The interesting outcome of these two theoretical studies is the observation of a magnetic threshold, also observed in the experiment. The closing of the gap by disorder has also been studied by *Laughlin* [3.34,35] in a three-parameter scaling theory. These theories are still in a state where a fair amount of *guesswork* is required as input [3.35]; further refinements are necessary.

A very interesting experiment on the energy gap has been performed recently by *Kukushkin* and *Timofeev* [3.36]. This is an alternative method to the thermal activation of the conductivity. They have measured the luminiscence spectra of the radiative recombination of two-dimensional electrons with photoexcited holes in Si-MOSFETs. In fact, the spectral position of the luminescence line, measured in the FQHE regime, is closely related to the chemical potential of interacting electrons. As we discussed earlier, we expect a discontinuity in the chemical potential at the filling fractions where the FQHE is observed. This discontinuity should manifest itself in the nonmonotonic behavior of the spectral position of the luminiscence line. In Fig. 3.24, we see that for a change of ν from 2.27 to 2.4 (around $\nu = \frac{7}{3}$), the luminiscence line shape is unaltered, but a nonmonotonic dependence of the spectral position is observed. Around $\nu = \frac{7}{3}$ (from $\nu = 2.32$ to $\nu = 2.35$) the luminiscence line has a doublet character (see Fig. 3.25). The observed doublet character of the luminescence spectrum indicates the presence of a gap in the spectrum of the incompressible fluid at a fractional filling.

In Fig. 3.26, the results for $\Delta E(\nu)$ measured for two Si-MOSFET samples with mobilities $\mu_e(T = 0.35\text{K}) = 4 \times 10^4 \text{cm}^2/\text{V.s}$ and $\mu_e(T = 0.35\text{K}) = 3 \times 10^4 \text{cm}^2/\text{V.s}$ with $B=8$ T are shown for filling factors around $\nu = \frac{7}{3}$, and $\frac{8}{3}$. Taking the valley degeneracy into account, these two filling factors actually describe $\frac{1}{3}$ and $\frac{2}{3}$ states. It is clear from Fig. 3.26 that for $\nu < \frac{7}{3}$ (and $\frac{8}{3}$), $\Delta E(\nu)$ is negative and has a minimum, while for $\nu > \frac{7}{3}$ (and $\frac{8}{3}$), $\Delta E(\nu)$ changes sign and reaches a maximum. This is explained by the fact that, in the process of recombination, the number of electrons is decreased by unity, thereby creating *three quasiholes*, while for $\nu > \frac{7}{3}$, *three quasiparticles* are created.

For $B=8$ T, the results for the creation energy for three quasiholes ($3\Delta_\text{h}$) and three quasiparticles ($3\Delta_\text{e}$) obtained from the above measurements are: $3\Delta_\text{e} = (4 \pm 0.3)K$ and $3\Delta_\text{h} = (3 \pm 0.3)K$ for $\nu = \frac{7}{3}$. The energy gaps $\Delta = \Delta_\text{e} + \Delta_\text{h}$ at $\nu = \frac{7}{3}$ and $\frac{8}{3}$ are close to the results obtained from thermal activation of conductivity.

There has been other recent interesting measurements of the activation gaps: *Boebinger* et al. measured the gap for $\nu = \frac{2}{5}$ and $\frac{3}{5}$ and have recently extended their work to other samples of different mobilities [3.31]. *Mendez*

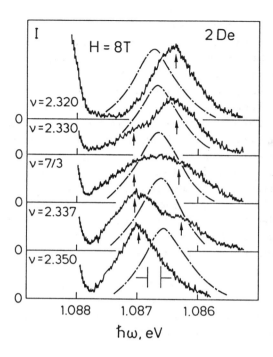

Fig. 3.25. The recombination spectra measured in the vicinity of $\nu = \frac{7}{3}$ at T=1.5 K and T=2.1 K (dashed line) for different values of ν: 2.320, 2.330, $\frac{7}{3}$, 2.337, 2.350 [3.36]

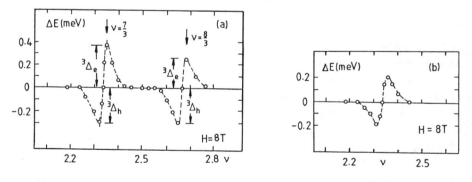

Fig. 3.26. Difference ΔE of spectral positions of the lines measured at T=1.5 K and T=2.1 K as a function of ν in the region of $\nu = \frac{7}{3}$ and $\nu = \frac{8}{3}$ for the two MOSFETs with mobility at T=0.35 K, (a) $\mu_e = 4 \times 10^4 \text{cm}^2/\text{V.s}$ and (b) $\mu_e = 3 \times 10^4 \text{cm}^2/\text{V.s}$ [3.36]

[3.37] has measured the activation energy at $\frac{1}{3}$ and $\frac{2}{3}$ for two-dimensional holes. *Guldner* et al. [3.38] reported microwave photoresistivity measurements in low density GaAs-heterojunctions and hinted that there might be nonzero activation energy for $B \lesssim 5T$, where *Boebinger* et al. observed a threshold. *Haug* et al. [3.39] have measured the activation energies for various filling fractions in a tilted magnetic field.

3.7 The Hierarchy: Higher Order Fractions

The theory of Laughlin for the ground state and the elementary excitations, is quite successful in describing the filling factors $\nu = \frac{1}{m}$ and $(1 - \frac{1}{m})$ with m an odd integer. However, experimental results for FQHE of filling factors such as, $\frac{2}{5}, \frac{2}{7}, \frac{3}{7}, \frac{4}{5}$ and the others described in the introduction clearly indicate that a nontrivial extension of Laughlin's theory is required to describe these states. The major efforts in this direction have been made using the hierarchial approach proposed by *Laughlin* [3.2,40], *Haldane* [3.41] and *Halperin* [3.42], where condensation of a finite density of quasiparticles is supposed to form the higher order states in the hierarchy. Another approach to the higher order filling fractions, based on the microscopic trial wave functions, has been carried out by Halperin and his collaborators.

According to *Laughlin* [3.40] the quasiparticle[5] motion can be understood by analogy with the electron motion. The quasiparticles behave much like electrons in the sense that their separations are quantized because of angular momentum conservation, but they are different from electrons in that the quantized separations are compatible with the *fractional statistics* as proposed by *Halperin* [3.42].

Following *Laughlin* [3.40], let us define the quasiparticle creation operators in the manner

$$S_{z_A}|m\rangle = e^{-\frac{1}{4}\sum_j |z_j|^2} \prod_i (z_i - z_A) \prod_{j<k} (z_j - z_k)^m \quad (3.49)$$

and

$$S^\dagger_{z_B}|m\rangle = e^{-\frac{1}{4}\sum_j |z_j|^2} \prod_i \left(2\frac{\partial}{\partial z_i} - z_B^*\right) \prod_{j<k} (z_j - z_k)^m \quad (3.50)$$

for a quasihole and a quasielectron respectively. In order to determine the two-quasiparticle eigenstates, Laughlin projected the Hamiltonian (2.20) of

---

[5] In this section, *quasiholes* and *quasielectrons* are collectively called the *quasiparticles*.

the many-electron system onto the set of states of the form $S_{z_A} S_{z_B}|m\rangle$ and diagonalized the projected Hamiltonian. The normalization integral is now given by

$$\langle m|S^\dagger_{z_B} S^\dagger_{z_A} S_{z_A} S_{z_B}|m\rangle = \int \ldots \int d^2 z_1 \ldots d^2 z_N \prod_{j<k} |z_j - z_k|^{2m} \prod_i |z_i - z_A|^2 \\ \times |z_i - z_B|^2 \, e^{-\frac{1}{2}\sum_l |z_l|^2}. \quad (3.51)$$

The integrand is readily recognized as the probability distribution of a classical plasma, $e^{-\mathcal{H}'_m}$ with the corresponding plasma Hamiltonian

$$\mathcal{H}'_m = -2m \sum_{j<k} \ln|z_j - z_k| + \frac{1}{2}\sum_l |z_l|^2 \\ -2\sum_i \left[\ln|z_i - z_A| + \ln|z_i - z_B|\right]. \quad (3.52)$$

In this case, (3.51) could be interpreted as the probability (to within a constant) of finding the *charge*-1 particles at z_A and z_B, if they are allowed to move around in the plasma. We therefore write

$$\langle m|S^\dagger_{z_B} S^\dagger_{z_A} S_{z_A} S_{z_B}|m\rangle = \frac{C}{|z_A - z_B|^{\frac{2}{m}}} e^{\frac{1}{2m}(|z_A|^2+|z_B|^2)} g_{22}(|z_A - z_B|) \\ = C e^{\frac{1}{2m}(|z_A|^2+|z_B|^2)} F\left[|z_A - z_B|^2\right] \quad (3.53)$$

where C is a constant and g_{22} is the radial distribution function for particles of *charge*-1. Laughlin found an approximate fit for F of the form

$$F\left[|z|^2\right] \simeq \frac{1}{4\pi m} \int d^2 z' \frac{1}{|z'|^{\frac{2}{3}}} e^{-\frac{1}{4}|z-z'|^2}. \quad (3.54)$$

Let us now choose the basis states

$$|z_A, z_B\rangle = e^{-\frac{1}{4m}(|z_A|^2+|z_B|^2)} S_{z_A} S_{z_B}|m\rangle. \quad (3.55)$$

The overlap matrix $\langle z_{A'}, z_{B'}|z_A, z_B\rangle$ is analytic in the variables $z_A, z_B, z^*_{A'}$ and $z^*_{B'}$. One could therefore analytically continue the normalization integral in the manner

$$\langle z_{A'}, z_{B'}|z_A, z_B\rangle = C e^{-\frac{1}{4m}(|z_A|^2+|z_B|^2+|z_{A'}|^2+|z_{B'}|^2)} \\ \times e^{\frac{1}{2m}(z^*_{A'} z_A + z^*_{B'} z_B)} F[(z_A - z_B)(z^*_{A'} - z^*_{B'})]. \quad (3.56)$$

The matrix elements of energy can be written similarly:

$$\langle z_{A'}, z_{B'}|\mathcal{H}|z_A, z_B\rangle = C\, e^{-\frac{1}{4m}(|z_A|^2+|z_B|^2+|z_{A'}|^2+|z_{B'}|^2)}$$
$$\times e^{\frac{1}{2m}(z_{A'}^* z_A + z_{B'}^* z_B)} E\left[(z_A - z_B)(z_{A'}^* - z_{B'}^*)\right], \qquad (3.57)$$

where E is fitted by the formula [3.40],

$$E\left[|z|^2\right] \simeq \frac{1}{4\pi m} \int d^2 z' \frac{1}{|z'|^{\frac{2}{m}}} e^{-\frac{1}{4m}|z-z'|^2} \left[\frac{(e/m)^2}{|z'|}\right] \qquad (3.58)$$

with the ground state energy taken to be zero.

These matrices are diagonalized by the states,

$$|n\rangle = \int\int d^2 z_A\, d^2 z_B\, e^{-\frac{1}{4m}(|z_A|^2+|z_B|^2)} (z_A^* - z_B^*)^n |z_A, z_B\rangle \qquad (3.59)$$

where n is an *even* integer, since the *Bose* representations for the quasiparticles are used. For the *Fermi* representation the basis states are to be chosen as $(z_A - z_B)|z_A, z_B\rangle$ and then n would be odd. The state $(z_A - z_B)|z_A, z_B\rangle$ is the electron-hole conjugate of the two-electron wave function for $m = 1$:

$$\psi(z_1, z_2) = \varphi_{z_A}(z_1)\varphi_{z_B}(z_2) - \varphi_{z_B}(z_1)\varphi_{z_A}(z_2) \qquad (3.60)$$

with

$$\varphi_{z_A}(z) = e^{-\frac{1}{4}|z|^2} e^{\frac{1}{2}z z_A^*} e^{-\frac{1}{4}|z_A|^2}. \qquad (3.61)$$

In the Fermi representation, the overlap matrix is,

$$\langle z_{A'}, z_{B'}|(z_{A'}^* - z_{B'}^*)(z_A - z_B)|z_A, z_B\rangle = C\, e^{-\frac{1}{4m}(|z_A|^2+|z_B|^2+|z_{A'}|^2+|z_{B'}|^2)}$$
$$\times e^{\frac{1}{2m}(z_{A'}^* z_A + z_{B'}^* z_B)} F^f\left[(z_A - z_B)(z_{A'}^* - z_{B'}^*)\right] \qquad (3.62)$$

where

$$F^f\left[|z|^2\right] = |z|^2\, F\left[|z|^2\right]. \qquad (3.63)$$

This is diagonalized by the wave function

$$|n+1\rangle = \int\int d^2 z_A\, d^2 z_B\, e^{-\frac{1}{4m}(|z_A|^2+|z_B|^2)} (z_A^* - z_B^*)^{n+1} (z_A - z_B)|z_A, z_B\rangle \qquad (3.64)$$

where $n + 1$ is odd. Laughlin then showed that this state is same as $|n\rangle$. Like the two-electron state in the lowest Landau level, the two-quasiparticle state does not depend on the repulsive potential between quasiparticles. It is also independent of the choice of basis. For more than two quasiparticles, *Laughlin* [3.43] found the fractional statistics representation of Halperin to be the most convenient one. This treatment of the quasiparticles will be discussed below.

According to *Haldane* [3.41,44], a Laughlin fluid state

$$\nu(m; \alpha_1, p_1; \alpha_2, p_2; \ldots ; \alpha_n, p_n)$$

of the excitations can be constructed from its parent state

$$\nu(m; \alpha_1, p_1; \alpha_2, p_2; \ldots ; \alpha_{n-1}, p_{n-1})$$

if $S_q = \frac{1}{2}(N_q - 1)p_n$ with the p_i as even integers (quasiparticles in this scheme obey Bose statistics); N_q and $(2S_q + 1)$ are the number of excitations and the degeneracy of the excitations of the parent state, respectively. The filling factor ν of the state is given as a continued fraction

$$\nu = \cfrac{1}{m + \cfrac{\alpha_1}{p_1 - \cfrac{\alpha_2}{p_2 - \cfrac{\alpha_3}{p_3 - \cdots}}}} \quad (3.65)$$

where $\alpha = +1$ (quasiholes) or -1 (quasiparticles). The construction of such a hierarchy is valid only to the extent that (i) short-range pair interaction pseudopotentials [defined in (2.60)] dominate the quasiparticle interactions, and (ii) these energies themselves are dominated by the collective excitation gap of the parent fluid. The iterative equations for the filling fractions in this hierarchial scheme are similar to those in the scheme of Halperin and are described later.

The excitation energy of the new fluid state is equivalent to that of a system consisting of N_q quasiparticles with fractional charge e_q obeying Bose statistics on the sphere of radius R and in a radial magnetic field $B_q = \hbar c S_q / e_q R^2$. The magnetic length for the quasiparticle state is $\ell_q = (\hbar c / e_q B_q)^{\frac{1}{2}}$. Zhang [3.45] showed how the excitation energy at any hierarchy level can be approximately related to the excitation energy for the $\frac{1}{m}$ state. Considering the Coulomb interaction between point particles of charge e_q, the excitation energy of the quasiparticles at any level of hierarchy is written as

$$E(\nu) = \left(\frac{e^2}{\epsilon \ell_0}\right) \frac{1}{Q_n^2 - 1} \left(\frac{p_n + 1}{Q_n}\right)^{\frac{1}{2}} f_F(p_n + 1) \quad (3.66)$$

where $f_F(m)$ is the excitation energy of the elementary Laughlin state for the electron system, and Q_n, Q_{n-1} are the denominators of the rational fillings of the new state and the parent state respectively. The above relation also preserves the electron-hole symmetry. The hierarchial scheme has been extended to the case of a system containing impurities [3.45].

The hierarchial scheme of *Halperin* [3.42] is very much in the spirit of Laughlin's theory. In this scheme, the quasiparticles, as mentioned above, are required to obey fractional statistics, i.e., when two particles are interchanged, the wave function of the system changes by a complex phase factor. If ν_t is a stable filling factor obtained at level t of the hierarchy, the low-lying energy states for filling factors near to ν_t can be described by the addition of a small density of quasiparticle excitations to the ground state at ν_t. There are two types of elementary excitations: *p*-excitations (particle like) and *h*-excitations (hole like), with charges $q_t e$ and $-q_t e$ respectively. Halperin then constructed a *pseudo* wave function (where the coordinates are for quasiparticles),

$$\psi(z_1,\ldots,z_{N_t}) = \prod_{i<j}(z_i - z_j)^{2p_{t+1}-\alpha_{t+1}/m_t} \exp\left[\sum_{i=1}^{N_t} -|q_t|^2/4\ell_0^2\right] \quad (3.67)$$

where z is the quasiparticle position, $\alpha_{t+1} = +1$ for the *p*-excitations and -1 for *h*-excitations, p_{t+1} is a positive integer, and N_t is the number of quasiparticles. The rational number m_t characterizes the fractional statistics of the quasiparticles.

Just like in Laughlin's theory for the ground state, the plasma analogy (2.26–29) is employed here to yield

$$m_{t+1} = 2p_{t+1} - \frac{\alpha_{t+1}}{m_t}. \quad (3.68)$$

The charge neutrality condition fixes the density of the plasma. A relation similar to (2.29) for the pseudo wave function shows that the number of quasiparticles in an area $2\pi\ell_0^2$ is $n_t = |q_t|/m_{t+1}$. As each quasiparticle has charge $\alpha_{t+1}q_t$, calculating the *electron* density in the new stable state, the filling factor is obtained from the following relation

$$\nu_{t+1} = \nu_t + \alpha_{t+1}\, q_t|q_t|/m_{t+1}. \quad (3.69)$$

Multiplying the pseudo wave function by the factor $\prod_k z_k$, $(k=1,\ldots,N_t)$, we find a *deficit* of $1/m_{t+1}$ quasiparticles at level t near the origin (see Sect. 3.1). This is a hole excitation at level $t+1$. In a similar manner, one can construct a *p*-excitation where one has an *excess* of $1/m_{t+1}$ quasiparticles at the origin. The iterative equation for q_t is then

$$q_{t+1} = \alpha_{t+1}\, q_t/m_{t+1}. \quad (3.70)$$

The starting values for the iterative equations, (3.68–70) are set to, $q_0 = m_0 = \alpha_1 = 1$. Then for any choice of the sequence $\{\alpha_t, p_t\}$, the iterative equations would provide a sequence of rational filling factors ν_t. The allowed values of ν_t may also be expressed as continued fractions in terms of the finite sequence $\{\alpha_t, p_t\}$.

As discussed by Halperin, quantized Hall steps will not, however, be observed for every rational ν. In fact, there exists a maximum value m_c such that, if at any stage of the hierarchy, the calculated m_t is greater than m_c, the quasiparticle density n_t will form a Wigner crystal (see Sect. 2.7), and it would be meaningless to continue from the corresponding electron density further in the hierarchy.

Assuming that at any stage of the hierarchy, the quasiparticles or quasiholes can be treated as point particles with pairwise Coulomb interactions, an estimate for the potential energy can be obtained as

$$E(\nu_{t+1}) \simeq E(\nu_t) + n_t\, \varepsilon_t^\pm(\nu_t) + n_t\, |q_t|^{\frac{5}{2}}\, E_{\rm pl}(m_{t+1}), \qquad (3.71)$$

where $E(\nu)$ is the energy per quantum of magnetic flux, $\varepsilon_t^\pm(\nu_t)$ is the energy required to add one p-excitation or h-excitation (gross energy[6]), and $E_{\rm pl}$ is the interpolation formula given in (2.37) [or (2.62)]. The factor $|q_t|^{\frac{5}{2}}$ reflects the smaller charge and larger magnetic length of the quasiparticles.

Let us consider the $\nu = \frac{2}{5}$ state. In this case, the parent state is $\nu = \frac{1}{3}$, and accurate Monte carlo results are available for the excitation energies at $\nu = \frac{1}{3}$. With $t = 1, p_1 = 2, \alpha_1 = 1$, the state $\frac{1}{3}$ is obtained. For the state $\nu = \frac{2}{5}$ we have $t = 2, p_2 = 1$ and $\alpha_2 = 1$. The iterative equation then yields, $m_1 = 3, q_1 = \frac{1}{3}, m_2 = \frac{5}{3}$, and $n_1 = \frac{1}{5}$. Making use of the result, $E(\frac{1}{3}) \simeq \frac{1}{3} E_{\rm pl}(3)$, we get the following relation for the energy

$$E\left(\frac{2}{5}\right) \simeq \frac{1}{5}\varepsilon_+\left(\frac{1}{3}\right) + \frac{1}{3} E_{\rm pl}(3) + \frac{1}{5}\left(\frac{1}{3}\right)^{\frac{5}{2}} E_{\rm pl}\left(\frac{5}{3}\right). \qquad (3.72)$$

The relation between ε_+ and the quasiparticle creation energy $\widetilde{\varepsilon}_{\rm p}$ (Sect. 3.1) is given in [3.4,41] as

$$\varepsilon_+\left(\frac{1}{3}\right) = \widetilde{\varepsilon}_{\rm p}\left(\frac{1}{3}\right) + \frac{1}{2} E_{\rm pl}(3). \qquad (3.73)$$

Dividing both sides of (3.72) by $\nu = \frac{2}{5}$, the energy per particle is written as

$$(E/N) \approx \frac{13}{12} E_{\rm pl}(3) + \frac{1}{2}\left(\frac{1}{3}\right)^{\frac{5}{2}} E_{\rm pl}\left(\frac{5}{3}\right) + \frac{1}{2}\widetilde{\varepsilon}_{\rm p}\left(\frac{1}{3}\right). \qquad (3.74)$$

---

[6] A detailed discussion of $\varepsilon_t^\pm$ is given by *Morf* and *Halperin* [3.4].

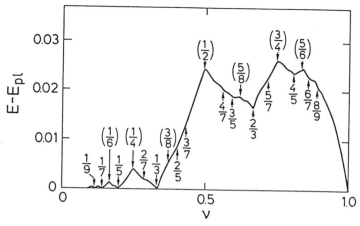

Fig. 3.27. Potential energy $E(\nu)$ per quantum of magnetic flux (in units of $e^2/\epsilon\ell_0$) vs the filling factor ν of the lowest Landau level. The smooth function $E_{\rm pl} = \nu\, E_{\rm pl}(m)$ has been subtracted from the result [3.42]

Using the Monte Carlo estimate, $\widetilde{\varepsilon}_{\rm p}(\frac{1}{3}) \approx 0.073\, e^2/\epsilon\ell_0$, the energy per particle is estimated to be $\sim -0.424\, e^2/\epsilon\ell_0$, which is quite close to the result $\sim -0.435\, e^2/\epsilon\ell_0$, obtained by *Yoshioka* et al. [3.12] by finite-size calculations in a periodic rectangular geometry.

Using a suitable iterative formula for the quasiparticle energies in (3.71), the energy versus density curve was generated by Halperin for various filling fractions shown in Fig. 3.27. In the figure, the stable fractions appear with downward pointing cusps. As *Laughlin* remarked [3.43], this curve has the interesting *fractal* property of being everywhere continuous but nowhere differentiable, with slope discontinuity at any point, which reflects the energy gap of the nearest allowed fraction.

The question of the quasiparticles statistics was also taken up by *Arovas* et al. [3.46]. With the help of the adiabatic theorem, they showed that an interchange of two quasiparticles (or quasiholes) along a counterclockwise loop gives a phase factor $e^{i\theta}$ with $\theta = \nu\pi$.

MacDonald et al. [3.47] proposed trial wave functions in terms of electron coordinates in each level of the hierarchy. Using an exact sum rule which is valid for any isotropic state in the lowest Landau level, they estimated the pair distribution functions and the energy of a hierarchy state by constructing a corresponding hierarchy of liquid structure functions. In Fig. 3.28, we present their results for the energies at various hierarchy states, together with the corresponding plasma energies [see (2.61)] and the CDW state energies in the HF approximation [3.48,49]. The essential difference between their results and those of Fig. 3.27 is, as these authors pointed out,

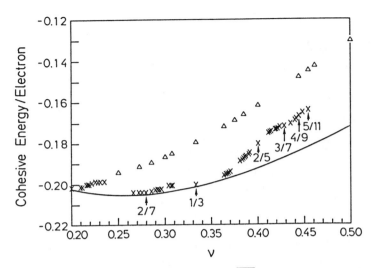

Fig. 3.28. Cohesive energy per electron ($E/N_e - \nu\sqrt{\pi/8}$) as a function of filling factor. The triangles are the CDW energies for a hexagonal lattice, the crosses are obtained using the hierarchy state pair-correlation function. The solid line is the interpolation formula for the plasma energy [3.47]

that the difference between the hierarchy state energy and the reference plasma energy tends to be larger when a condensate occurs in the quasi-electrons rather than the quasiholes. Following this line of approach, *MacDonald* and *Murray* [3.50] constructed trial wave functions for up to eight electrons and obtained results for the ground state and excitation energies for states associated with $\nu = \frac{1}{3}, \frac{2}{7}$, and $\frac{2}{5}$.

Finally, the hierarchial scheme described above has been employed by *Zhang* and *Chakraborty* [3.51] to study the condensation of the spin-reversed quasiparticles. The trial wave function for the quasiparticle excitations of the $\nu = \frac{1}{m}$ states with one spin-reversed electron is given in (3.36). For a number n_t of spin-1 quasiparticles in the Laughlin state at $\nu = \frac{1}{m}$, the trial wave function is written

$$\psi(z_1,\ldots,z_{N_e}) = \prod_{i<j}^{N_e}(z_i - z_j)^{m-1+\delta_{\sigma_i\cdot\sigma_j}} \exp\left[-\frac{1}{4\ell_0^2}\sum_{i=1}^{N_e}|z_i|^2\right], \quad (3.75)$$

where $\sigma_i = -1$ for $i = 1,\ldots,N_t$, and $\sigma_i = 1$ for $i = N_t+1,\ldots,N_e$, z_i is, as usual, the complex coordinate of the ith electron. Let us consider the following two cases, (a) $n_t = 1$ and (b) $n_t = \frac{1}{2}N_e$. In the former case, we have the one spin-1 quasiparticle state, and (3.75) reduces to the state (3.36). In case (b), we have the spin unpolarized state at $\nu = 2/(2m-1)$, and (3.75) reduces to the state (2.63) discussed in Sect. 2.5.

As for the microscopic trial wave function approach, several trial wave functions for higher order filling factors were suggested by *Halperin* [3.3]. The most extensive numerical calculations reported so far, are for $\nu = \frac{2}{5}$. There are also results for $\frac{2}{7}$, $\frac{2}{9}$ and for $\frac{2}{3}$. The trial wave function used in these studies is written as [3.6]

$$\psi = \mathcal{A}\widetilde{\psi}$$

$$\widetilde{\psi} = \left[\prod_k e^{-|z_k|^2/4\ell_0^2}\right]\left[\prod_{k<l}(z_k - z_l)^s\right]\left[\prod_i (z_{2i-1} - z_{2i})^t\right]\left[\prod_{i<j}(Z_i - Z_j)^{2u}\right]$$

(3.76)

where s, t and u are required to be integers with $s > 0, u \geq 0, s - t > 0$, and $s - t$ odd. Furthermore, u or t or both required to be > 0. In (3.76), k and l run from 1 to N_e, while i and j run from 1 to $\frac{1}{2}N_e$ and $Z_i = \frac{1}{2}(z_{2i} + z_{2i-1})$ is the center of gravity of the ith pair. The corresponding filling factor is given by $\nu = 2/(2s + u)$. The motivation behind constructing such a trial wave function has been discussed by *Halperin* [3.3].

For the $\frac{2}{5}$ state ($s = 2, u = t = 1$), the Monte Carlo results for a disk geometry give an estimate of potential energy per electron of $-0.414 \pm 0.002\, e^2/\epsilon\ell_0$. This energy is much higher than the value of $\sim -0.435\, e^2/\epsilon\ell_0$ obtained from exact diagonalization of systems with up to eight electrons in a periodic rectangular geometry. The trial wave function for $\frac{2}{5}$ is therefore not a good approximation for the true ground state.

For the $\frac{2}{7}$ state ($s = 3, t = 0, u = 1$), a reasonably good result was obtained, $E/N_e \approx -0.377\, e^2/\epsilon\ell_0$, as compared to $E/N_e \approx -0.385\, e^2/\epsilon\ell_0$ for a four-electron system in a periodic rectangular geometry. However, in this case the accuracy required for the calculations to be useful is not achieved [3.4].

For the $\frac{2}{3}$ state ($s = u = 1, t = 0$), the trial wave function provides the result $E/N_e \approx -0.509\, e^2/\epsilon\ell_0$. The $\frac{2}{3}$ state energy can also be computed from Laughlin's $\frac{1}{3}$ state via the electron-hole symmetry relation (Sects. 2.1,2):

$$\nu u(\nu) = (1-\nu)u(1-\nu) + \sqrt{\frac{\pi}{8}}\frac{e^2}{\epsilon\ell_0}(1 - 2\nu)$$

(3.77)

where $u(\nu)$ is the potential energy per electron in the state. One then obtains $E_L(\frac{2}{3})/N_e \approx -0.518\, e^2/\epsilon\ell_0$ which is lower than the trial wave function result.

An alternative wave function has been explored in the spherical geometry for the $\frac{2}{5}$ and $\frac{2}{7}$ state [3.19,52] and for the $\frac{2}{9}$ state [3.18]. The estimate for the $\frac{2}{5}$ state in the thermodynamic limit is given by $E/N_e \approx -0.4303 \pm 0.0030\, e^2/\epsilon\ell_0$ [3.19].

Experimentally, there are indications that the FQHE occurs in multiple series p/q, with fractions of odd denominators [3.53]. Furthermore, at a given temperature, the $\frac{1}{3}$ effect is always the best developed, followed by $\frac{2}{5}, \frac{3}{7}$, and $\frac{4}{9}$. The $\frac{2}{3}$ sequence ($\frac{3}{5}, \frac{4}{7}, \frac{5}{9}$) shows a similar behavior. As the magnetic field strength varies by only $\approx 25\%$ within each sequence, the difference in field strength is probably not the reason for the decrease in strength of the effects within a sequence. Further work on the hierarchial models discussed in this section is required to provide a better explanation of these experimental findings.

Note added in Proof: After the review was completed in January 1988, some new and interesting experimental results came to our attention:

Clark et al. [R. G. Clark, J. R. Mallett, S. R. Haynes, J. J. Harris C. T. Foxon: Phys. Rev. Lett. **60**, 1747 (1988)] reported a systematic study of $\sigma_{xx}^c = \sigma_{xx}(\frac{1}{T} = 0)$ for a range of GaAs-heterojunctions and $\frac{p}{q}$ states with $q = 3, 5, 7$, and 9 obtained from

$$\sigma_{xx}^c = \frac{\rho_{xx}^c}{(\rho_{xx}^c)^2 + \rho_{xy}^2} = \frac{\rho_{xx}^c}{(\rho_{xx}^c)^2 + \left[\left(\frac{q}{p}\right)\frac{h}{e^2}\right]^2}$$

which is valid at exactly $\nu = \frac{q}{p}$, and $\rho_{xx} = \rho_{xx}^c\, e^{-\Delta/k_B T}$ with Δ the energy gap. They found that, within experimental errors, σ_{xx}^c is constant for $\frac{p}{q}$ fractions of the same q and scales as $\frac{1}{q^2}$. This observation provides for the first time, an experimental probe of e^* and confirms that $e^* = \pm\frac{e}{q}$.

Mallett et al. [J. R. Mallett, R. G. Clark, R. J. Nicholas, R. Willett, J. J. Harris and C. T. Foxon: Phys. Rev. **B38**, 2200 (1988)] recently reported an experimental study of the $\nu = \frac{1}{5}$ ground state and determined the activation energy $\Delta_{\frac{1}{5}}$. They also reported observation of weak ρ_{xx} minima at $\nu = \frac{2}{9}, \frac{2}{11}$, thereby providing experimental confirmation of the $\frac{1}{5}$ hierarchy, which is separate from the sequence of states derived from $\nu = \frac{1}{3}$.

Clark et al. [R. G. Clark: private communication to T. C., and Proc. of 19th Int. Conf. on Physics of Semiconductors, Warsaw, Aug. 1988, and to be published] have also studied experimentally the ground state spin polarization in the lowest Landau level. Results for some of the filling fractions presented in Table 2.1 are verified in this experiment. Finally, there are also experimental indications for the presence of spin-reversed quasiparticles, discussed in Sect. 3.3.

4. Collective Modes: Intra-Landau Level

In this chapter, we discuss the intra-Landau level collective excitations occurring when the lowest Landau level is fractionally filled. The calculations are mostly focused on the filling factor $\frac{1}{3}$. A few results are available for $\frac{2}{5}$ and are also discussed. The finite-size studies (in spherical geometry [4.1] as well as in the periodic rectangular geometry [4.2]) are very effective in determining the low-lying collective modes. A Laughlin-type approach with the HNC scheme has not been very successful so far because of the inability of the HNC approach to obtain the correct quasiparticle excitation energy (see preceding chapter, and also [4.3]). Based on the Laughlin wave function for the ground state, the single-mode approximation has been quite successful in bringing out a physical picture for the collective mode. These topics are discussed in detail below.

4.1 Finite-Size Studies: Spherical Geometry

The calculation of the low-lying excitation spectrum for the incompressible fluid state at $\nu = \frac{1}{3}$ in finite-size systems was initiated by Haldane [4.1,2]. We have discussed earlier their results for the ground state and the quasiparticle-quasihole states obtained in a spherical geometry. In this geometry, the full energy-level spectrum for a seven-electron system at $\nu = \frac{1}{3}$ in the lowest Landau level is shown in Fig. 4.1. The figure contains several interesting features. The ground state appears to be non-degenerate with angular momentum $L = 0$, and is well separated from the other energy levels. The set of low-lying excitations for increasing value of L may be identified as a branch of neutral elementary excitations.

In the spherical geometry, the states of a neutral particle are described by spherical harmonic wave functions. The spectrum for neutral excitation is plotted in Fig. 4.2, as a function of the effective wave number $k = L/R$, where L is the total angular momentum and R is the radius of the sphere. The lowest energy excitations clearly show a collective behavior, well separated from the higher energy states. The excitation gap has a minimum at $k\ell_0 \approx 1.4$. The large L (large k) limit is identified as a well separated quasiparticle-quasihole pair. The mean square separation (chord distance)

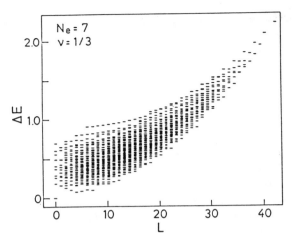

Fig. 4.1. Complete excitation spectrum of 1656 multiplets (50388 states) of a seven-electron system at $\nu = \frac{1}{3}$ in a spherical geometry with the lowest Landau level Coulomb interaction [4.1]

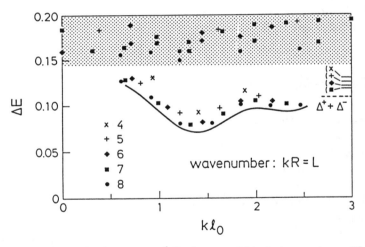

Fig. 4.2. Low-lying excitations at $\nu = \frac{1}{3}$ for four- to eight-electron systems. The full line is a guide to the eye. The estimate of $E_g = \Delta^+ + \Delta^-$ for various system size are also given [4.1]

is given as $2RL/N = k\ell_0^2/\nu$. At large k, the excitation energy of the *quasiexciton* must approach the value of E_g as discussed in the preceding chapter. Numerical calculations for the collective excitation spectrum in the spherical geometry have also been performed by *Fano* et al. [4.4] at $\nu = \frac{1}{3}$ for a nine-electron system and at $\nu = \frac{1}{5}$ for a seven-electron system. Their results for the lowest energy excitations are shown in Fig. 4.3.

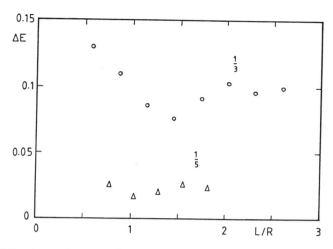

Fig. 4.3. Collective mode at $\nu = \frac{1}{3}$ (for $N_e = 9$ and plotted as circles) and at $\nu = \frac{1}{5}$ (for $N_e = 7$ and plotted as triangles) [4.4]

4.2 Rectangular Geometry: Translational Symmetry

In the case of a periodic rectangular geometry, the major contribution to calculating the collective excitation spectrum was also by *Haldane* [4.2]. The earlier works by *Su* [4.5,6] on the excitation spectrum in a periodic rectangular geometry were based on the formalism developed by Yoshioka et al. (Sect. 2.1), which gives a q-fold degenerate ground state at $\nu = p/q$. In Laughlin's theory, as well as in the spherical geometry results, the ground state, however, is nondegenerate. This q-fold ground state degeneracy was considered by many authors as an intrinsic feature to the FQHE [4.7–9]. In his work, Haldane pointed out that the formalism described in Sect. 2.1 employs essentially one-particle symmetry analysis. Introducing the translational symmetry into the system, Haldane then found that the q-fold ground state degeneracy could be identified as a center-of-mass degeneracy common to all states and is without any physical significance. It is present, irrespective of whether the system is an incompressible fluid state or not.

The most interesting outcome of Haldane's analysis was that, at rational values of ν, the states could be characterized by a two-dimensional wave vector k, and hence the collective excitation spectrum could be calculated in the periodic rectangular geometry. In the following, we describe in detail the translational symmetry analysis of a two-dimensional many-electron system in a magnetic field.

As we recall, in Sect. 2.1 the finite-size studies were carried out in an infinite lattice with a rectangular geometry. The electrons in one cell of the lattice have identical mirror images in all other cells. This infinite repetition will introduce a symmetry which can be employed to classify the eigenstates of the Hamiltonian. This classification has a further advantange that the size of the matrix to be numerically diagonalized is reduced.

In the absence of a magnetic field, the symmetry analysis would be simple: we would have a translational group in the periodic lattice and the eigenstates could be labelled by the wave vectors in the inverse lattice. The physical interpretation of these quantum numbers would of course be the momentum. The presence of the magnetic field, however, slightly complicates the classification scheme. To proceed we could apply the well-known apparatus of group theory which would lead us to the so called *ray representation* of the magnetic translation group [4.2,10–12]. The reason why we do not obtain an ordinary representation is that the symmetry operations of the group obey a non-commutative algebra where the product of operators is an operator of the same group only to within a phase factor. We will take a more direct approach, however. Our aim is to find a complete set of operators which commute with the Hamiltonian and can therefore be simultaneously diagonalized with it. To this end we will closely follow the work of *Haldane* [4.2].

We start with the Hamiltonian operator in configuration space

$$\mathcal{H} = \frac{1}{2m} \sum_j \Pi_j^2 + \frac{1}{2} \sum_{i \neq j} V(r_i - r_j) \tag{4.1}$$

where the momentum in the presence of the vector potential A is given by

$$\Pi_j = -i\hbar \nabla_j - eA(r_j). \tag{4.2}$$

Since we are working in rectangular geometry the most natural choice for the vector potential is to use the Landau gauge. We therefore consider the case, $A = Bx\hat{y}$. The magnetic field B is then perpendicular to our two-dimensional configuration space. The first step is to find an operator which commutes with the Hamiltonian. It is easy to verify that the quantity

$$K_j = -p_j - eBy_j\hat{x} \tag{4.3}$$

is such an operator. We now want an operator which corresponds to the translation operator in the non-magnetic field case and can thus be used to construct the momentum eigenstates of the system. We therefore proceed as in the normal system and define the *magnetic translation operator*

$$T(L) = \exp\left\{-\frac{i}{\hbar} L \cdot K\right\}. \tag{4.4}$$

When we make use of the operator relation

$$e^A e^B = e^{\frac{1}{2}[A,B]} e^{A+B},$$

provided that $[A,[A,B]] = [B,[A,B]] = 0$, the operator T can be split into two parts

$$T_j(\boldsymbol{L}) = \exp\left\{\frac{i}{\ell_0^2}(L_x y_j - \frac{1}{2}L_x L_y)\right\} t_j(\boldsymbol{L}) \qquad (4.5)$$

where t_j is the ordinary translation operator.[1] Using this relation it is easy to show that the magnetic translation operators obey the algebra given by

$$T_j(\boldsymbol{L}_1)T_j(\boldsymbol{L}_2) = \exp\left\{\frac{i}{2\ell_0^2}\widehat{\boldsymbol{z}} \cdot \boldsymbol{L}_1 \times \boldsymbol{L}_2\right\} T_j(\boldsymbol{L}_1 + \boldsymbol{L}_2), \qquad (4.6)$$

which is a gauge independent result. From (4.6) it follows that,

$$T_j(\boldsymbol{L}_1)T_j(\boldsymbol{L}_2) = \exp\left\{\frac{i}{\ell_0^2}\widehat{\boldsymbol{z}} \cdot \boldsymbol{L}_1 \times \boldsymbol{L}_2\right\} T_j(\boldsymbol{L}_2)T_j(\boldsymbol{L}_1). \qquad (4.7)$$

It is therefore impossible to remove the phase factor by redefinition of the magnetic operators. The magnetic translations thus form a ray representation of the translation group and cannot be transformed into a vector representation. As a special application of (4.6), we get the quantization rule for the number N_s of magnetic flux quanta passing through a unit cell as

$$L_x L_y = 2\pi \ell_0^2 N_s, \qquad (4.8)$$

when we let the translation vectors $\boldsymbol{L}_1$ and $\boldsymbol{L}_2$ be the unit vectors of the lattice and require that circling the unit cell gives us the identity operator, i.e.

$$T_j(L_x\widehat{\boldsymbol{x}})T_j(L_y\widehat{\boldsymbol{y}})T_j(-L_x\widehat{\boldsymbol{x}})T_j(-L_y\widehat{\boldsymbol{y}}) = 1. \qquad (4.9)$$

From (4.9) we obtain,

$$\frac{1}{\ell_0^2}\widehat{\boldsymbol{z}} \cdot \boldsymbol{L}_1 \times \boldsymbol{L}_2 = 2\pi N_s. \qquad (4.10)$$

The phase factor in (4.6) is thus related to the number of flux quanta passing through the cell. Due to the periodic structure of the configuration space the physical quantities must be invariant under magnetic translations by the

---

[1] In the case of the symmetric gauge vector potential, $\boldsymbol{A} = \frac{1}{2}\boldsymbol{B} \times \boldsymbol{r}$, the magnetic translation operator can easily be evaluated to be,

$$T_j(\boldsymbol{L}) = \exp\left\{\frac{i}{2\ell_0^2}(\widehat{\boldsymbol{z}} \times \boldsymbol{L}) \cdot \boldsymbol{r}_j\right\} t_j(\boldsymbol{L}).$$

amount of any lattice vector. Thus the physical states must be constructed from the eigenstates of these operators and the physical operators from the gauge invariant products of these same eigenstates. By applying a translation which does not correspond to a lattice vector we merely move the representation of our system to another Hilbert space.

Thus far we have treated only single particle operators. To proceed to the many-body system it is found to be useful to define quantities p and q such that the number of electrons N_e and flux quanta N_s in the unit cell can be expressed in the forms $N_e = pN$ and $N_s = qN$ respectively. Here N is the highest common divisor of N_e and N_s. The filling factor ν in terms of these quantities is $\nu = p/q$. The center of mass (CM) translation operator $\bar{T}$ which moves every particle by the same vector a is clearly given by

$$\bar{T}(a) = \prod_i T_i(a). \tag{4.11}$$

We now require that the CM translation will not affect the single particle states which are labelled by the eigenvalues of the operators T_i. The most general translation satisfying this requirement is found to be of the form [4.2],

$$a = \frac{1}{N_s} L_{mn} \tag{4.12}$$

where L_{mn} is a lattice vector $[mL_x\hat{x} + nL_y\hat{y}]$. Other forms of translation will change the Hilbert space of the system.

Physically, the the overall motion of the system is not very interesting. We therefore split the translation of the particle i into two parts: the CM motion and the relative motion. We define the *relative translation operator* $\widetilde{T}_i$ acting on particle i in such a way that the motion of that particle is compensated by the movement of all the other particles in the opposite direction thus leaving the CM of the system untouched. We then have,

$$\widetilde{T}_i(a) = \prod_j T_i(a/N_e)T_j(a/N_e). \tag{4.13}$$

Using the algebra (4.6) it is straightforward to verify that the system is invariant under the application of this operator to all the particles, i.e.

$$\prod_i \widetilde{T}_i(a) = 1. \tag{4.14}$$

The decomposition of the single particle translation into the CM and relative parts can easily be found to be

$$T_i(a) = T(a/N_e)\widetilde{T}_i(a). \tag{4.15}$$

The relative translation operators will be our fundamental operators providing the quantum numbers we are seeking for the classification of the states of the system. It is thus important to find all operators which leave the Hilbert space invariant and commute with each other. These operators can therefore be simultaneously diagonalized—together with the Hamiltonian. The eigenvalues of the operators of this set will give us a complete set for labelling the many-particle states. In other words our task is to find a maximum set $\mathcal{M}$ of translation vectors for which the commutation relation

$$\left[\widetilde{T}_i(\boldsymbol{a}), \widetilde{T}_j(\boldsymbol{b})\right] = 0 \tag{4.16}$$

holds for any particle i and j and for all vectors $\boldsymbol{a}, \boldsymbol{b} \in \mathcal{M}$. To satisfy the requirement for the invariance of the Hilbert space, these operators must also commute with the operators $T_k(\boldsymbol{L}_{mn})$. Explicit calculations show that this maximum set is given by

$$\mathcal{M} = \{p\boldsymbol{L}_{mn}\}. \tag{4.17}$$

Having now a suitable set of operators at our disposal we still have to give a physical interpretation to their eigenvalues. To this end, we write the eigenvalues of the operator $\widetilde{T}_i(\boldsymbol{a})$ in the form $e^{i\boldsymbol{k}\cdot\boldsymbol{a}}$ in analogy with the normal translation operators. We now make the physically plausible statement that the application of the normal momentum operator $\sum_i e^{i\boldsymbol{Q}\cdot\boldsymbol{r}_i}$ to a many-particle state will increase its momentum by an amount $\boldsymbol{Q}$ provided $\boldsymbol{Q}$ is an allowed wave vector, i.e. a vector in the inverse lattice. It is straightforward to verify the relation

$$\widetilde{T}_i(p\boldsymbol{L}_{mn})\left(\sum_j e^{i\boldsymbol{Q}\cdot\boldsymbol{r}_j}\right) = e^{i\boldsymbol{Q}\cdot p\boldsymbol{L}_{mn}/N_e}\left(\sum_j e^{i\boldsymbol{Q}\cdot\boldsymbol{r}_j}\right)\widetilde{T}_i(p\boldsymbol{L}_{mn}) \tag{4.18}$$

from which we can deduce that the eigenvalues of the relative translation operator can be written in the form $e^{2\pi i(ms+nt)/N}$ where s and t are integers. As a special case we can take the vector $\boldsymbol{L}_{mn}$ to be $L_x\widehat{\boldsymbol{x}}$ and $L_y\widehat{\boldsymbol{y}}$ in turn and find that the spectrum of $\widetilde{T}_i$ consists of N^2 points in the inverse lattice.[2] From (4.18) we can also find the connection to the true physical momentum:

$$k\ell_0 = \sqrt{\frac{2\pi}{N_s\lambda}}\left(s - s_0, \lambda(t - t_0)\right). \tag{4.19}$$

---

[2] It is interesting to note that, in the case of *irrational* filling factors which are obtained as the limit of a sequence where N_e and N_s have the common divisor $N = 1$, the two-dimensional reciprocal space cannot be constructed. In this case, the symmetry analysis of Yoshioka et al. in Sect. 2.1 is complete (see [4.2]).

Here we have denoted by λ the aspect ratio L_x/L_y and by (s_0, t_0) the quantum numbers corresponding to zero momentum. This particular state can be identified by its symmetry properties. In reciprocal space the $\boldsymbol{k} = 0$ state has the highest symmetry and should therefore remain invariant under all symmetry operations. Thus the eigenvalues of the relative translation operator for this particular state must satisfy

$$\widetilde{T}_i(pL_x\hat{\boldsymbol{x}}) = \widetilde{T}_i(pL_y\hat{\boldsymbol{y}}) = \widetilde{T}_i(pL_y\hat{\boldsymbol{y}} - pL_x\hat{\boldsymbol{x}}). \tag{4.20}$$

After a little algebra we find that

$$e^{2\pi i s_0/N} = e^{2\pi i t_0/N} = (-1)^{pq(N_e-1)}. \tag{4.21}$$

This relation will uniquely determine the $\boldsymbol{k} = 0$ state in the Brillouin zone of the reciprocal lattice. If for example the number of electrons N_e is odd then the state for which $(s,t) = (0,0)$ also has $\boldsymbol{k} = 0$. For an even number of electrons the analysis is only slightly more involved. The incompressible fluid state is now characterized as states where the relative part of the Hamiltonian has a nondegenerate ground state at $\boldsymbol{k} = 0$ and a finite gap for all excitations.

Having completed the symmetry analysis we will apply previous results to the construction of basis states suitable for numerical diagonalization of the Hamiltonian matrix. These states are formed from the single particle eigenstates,

$$\begin{aligned}\phi_{Kj}(\boldsymbol{r}) = C_K \sum_k \exp\left[\frac{i}{\ell_0^2}(X_j + kL_x)y - \frac{1}{2\ell_0^2}(X_j + kL_x - x)^2\right] \\ \times H_K\left[\frac{1}{\ell_0}(X_j + kL_x - x)\right]\end{aligned} \tag{4.22}$$

of the Hamiltonian operator $\mathcal{H} = \frac{1}{2m}\Pi^2$. Here C_K is the normalization constant, $X_j = \frac{2\pi\ell_0^2}{L_y}j$, and H_K is the Hermite polynomial. The quantum number K labels the Landau level and the momentum label j takes values $0, 1, \ldots$ (mod N_s). A direct calculation shows that the states (4.22) are also eigenstates of the translation operators $T(L_{mn})$ defined in (4.5) with eigenvalues 1. We denote by $|j_1, j_2, \ldots, j_{N_e}\rangle$ the state of N_e electrons constructed from the single particle states (4.22). For simplicity we have omitted the Landau level index as well as the spin label of the electrons. Implicitly, however, we assume that these labels are included in the indices j_i. It is a simple matter to show that these product states are eigenstates of a subset of the relative translation operators:

$$\widetilde{T}_i(pnL_y\hat{y})|j_1, j_2, \ldots, j_{N_e}\rangle = \exp\left\{2\pi i \frac{n}{N} t\right\} |j_1, j_2, \ldots, j_{N_e}\rangle. \qquad (4.23)$$

Here t is the sum of the individual momenta of all particles, $t = \sum_i j_i$ (mod N). On the other hand translations in the x-direction will not leave these states invariant. They are instead mapped to another state, namely

$$\widetilde{T}_i(pmL_x\hat{x})|j_1, j_2, \ldots, j_{N_e}\rangle = |j_1 - qm, j_2 - qm, \ldots, j_{N_e} - qm\rangle. \qquad (4.24)$$

Using this last relation it is straightforward to construct superpositions which are eigenstates of an arbitrary relative translation operator. We denote by $\mathcal{L}$ the minimum set of all states $|j_1, j_2, \ldots, , j_{N_e}\rangle$ for which the total momentum is t and every member $|j_1, j_2, \ldots, j_{N_e}\rangle$ of the set $\mathcal{L}$ is related to every other member $|j_1', j_2', \ldots, j_{N_e}'\rangle$ of the same set by

$$|j_1', j_2', \ldots, j_{N_e}'\rangle = |j_1 - qk, j_2 - qk, \ldots, j_{N_e} - qk\rangle \qquad (4.25)$$

where k is an arbitrary integer. The number of elements $|\mathcal{L}|$, in this set, is at most N since the momenta j_i are defined (mod N_s) and $qN = N_s$. The above relation can also be viewed as an equivalence relation which divides all states with a given t into equivalence classes $\mathcal{L}$. The number of these classes is roughly the fraction $1/N$ of the total number of the states. It is clear from the construction that all states in a set $\mathcal{L}$ are mapped by (4.24) to states in the same set. The requirement of the minimality of the set $\mathcal{L}$ also guarantees that $\mathcal{L}$ does not contain any proper subset with the same properties. Applying the operator $\widetilde{T}_i(L_{mn})$ to the state

$$|(0,t);\mathcal{L}\rangle = \sum_{k=0}^{|\mathcal{L}|-1} |j_1 - qk, j_2 - qk, \ldots, j_{N_e} - qk\rangle, \quad |j_1, j_2, \ldots, j_{N_e}\rangle \in \mathcal{L}$$
$$(4.26)$$

reveals that it is indeed an eigenstate of this operator with the eigenvalues $e^{2\pi i n t/N}$. The remaining eigenstates can be found with the help of the relation (4.18) which we use to create momentum in the x-direction. Thus we finally arrive to the complete set of normalized states

$$|(s,t);\mathcal{L}\rangle = \frac{1}{\sqrt{|\mathcal{L}|}} \sum_{k=0}^{|\mathcal{L}|-1} \exp\left\{\frac{2\pi i s}{N} k\right\} |j_1 - qk, j_2 - qk, \ldots, j_{N_e} - qk\rangle \qquad (4.27)$$

which belong to the relative momentum (4.19) and are appropriate for the numerical finite-size studies.

The procedure to construct the basis states is now summarized as follows: Let us start with the filling fraction $\nu = p/q$, where the integers p and q have no common factors other than 1. Let us now select the number of electrons such that $N_e = pN$ for some integer N. The number of flux quanta would then be, $N_s = qN$. Next we choose the momentum label t, $0 \leq t < N$, in the y-direction and construct the set of all states $|j_1, j_2, \ldots, j_{N_e}\rangle$ for which $\sum j_i = t \pmod{N_s}$. We then partition this set into equivalence classes in such a way that each member $|j_1, j_2, \ldots, j_{N_e}\rangle$ of an equivalence class $\mathcal{L}$ is related to any other member $|j'_1, j'_2, \ldots, j'_{N_e}\rangle$ of the same class by (4.25). We now select the momentum label s, $0 \leq s < N$, in the x-direction and map each class by (4.27) to a momentum eigenstate corresponding to the total momentum k given by (4.19).

With the basis states constructed as above, the excitation spectrum is now calculated for a system containing a finite number of spin-polarized electrons [4.2,13]. The many-electron Hamiltonian, (2.14–16) is rewritten as

$$\mathcal{H} = \sum_j K\hbar\omega_c a^\dagger_{Kj} a_{Kj} + \sum_{\substack{j_1 j_2 \\ j_3 j_4}} \sum_{\substack{K_1 K_2 \\ K_3 K_4}} \mathcal{A}_{K_1 j_1 K_2 j_2 K_3 j_3 K_4 j_4} a^\dagger_{K_1 j_1} a^\dagger_{K_2 j_2} a_{K_3 j_3} a_{K_4 j_4}$$

(4.28)

where K is the Landau level index and,

$$\mathcal{A}_{K_1 j_1 K_2 j_2 K_3 j_3 K_4 j_4} = \delta'_{j_1 + j_2, j_3 + j_4} \mathcal{F}_{K_1 K_2 K_3 K_4}(j_1 - j_4, j_2 - j_3), \qquad (4.29)$$

$$\begin{aligned}
\mathcal{F}_{K_1 K_2 K_3 K_4}(j_a, j_b) = &\frac{1}{2ab} \sideset{}{'}\sum_q \sum_{k_1} \sum_{k_2} \delta_{q_x, 2\pi k_1/a} \delta_{q_y, 2\pi k_2/b} \delta'_{j_a k_2} \\
&\times \frac{2\pi e^2}{\epsilon q} \left[\frac{8 + 9(q/b) + 3(q/b)^2}{8(1 + q/b)^3}\right] \mathcal{B}_{K_1 K_4}(q) \mathcal{B}_{K_2 K_3}(-q) \\
&\times \exp\left(-\tfrac{1}{2} q^2 \ell_0^2 - 2\pi i k_1 j_b / N_s\right),
\end{aligned}$$

(4.30)

$$\mathcal{B}_{K_1 K_2}(q) = \begin{cases} 1 & K_1 = K_2 = 0 \\ -\dfrac{iq_x + q_y}{\sqrt{2}} \ell_0 & K_1 = 0,\ K_2 = 1 \\ -\dfrac{iq_x - q_y}{\sqrt{2}} \ell_0 & K_1 = 1,\ K_2 = 0 \\ \left(1 - \tfrac{1}{2} q^2 \ell_0^2\right) & K_1 = K_2 = 1 \end{cases}.$$

(4.31)

In (4.30), the finite-thickness correction is also included, with the Fang-Howard variational parameter b (see Sect. 2.6). In the following, we consider the pure two-dimensional case ($b = \infty$) only.

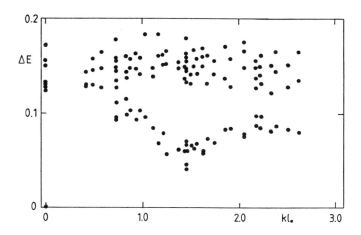

Fig. 4.4. Low-lying excitation energies for finite electron systems at $\nu = \frac{1}{3}$

In Fig. 4.4 we present the results for the density wave mode obtained for four- to seven-electron systems in the lowest Landau level. Only the three lowest excitation energies are shown. The spectrum is in fact, a function of the two-dimensional vector k. In the figure we consider only the absolute value of k. The ground state is obtained at $k = 0$, as expected. The lowest energy excitations are separated from the ground state by a large gap, which reflects the incompressible nature of the system and they clearly show a collective behavior with a minimum at finite $k\ell_0$ which, as we have seen for the spherical geometry, is a characteristic feature of the excitation spectrum. For small $k\ell_0$, the modes are not very well defined, as they are close to the continuum of the higher energy states. The numerical calculations were done by Yoshioka with a value of aspect ratio of $N_e/4$, but qualitatively similar results were also obtained earlier by Haldane for a six-electron system in a square geometry. The results are also very similar to the spectrum obtained in the spherical geometry discussed above.

The excitation spectrum as obtained above, can also be used to estimate the quasiparticle–quasihole energy gap, E_g. If we identify the lowest lying excitations as the quasiexcitons, E_g would be the asymptotic value of the lowest lying collective dispersion $E(k)$ obtained above numerically. As noted by *Kallin* and *Halperin* [4.14], for large values of $k\ell_0$, the quasiexcitons comprise a quasiparticle and a quasihole separated by a large distance, $|\Delta r| = k\ell_0^{*2} = k\ell_0^2 m$, where ℓ_0^* is the effective magnetic length for a particle of charge $e^* = \pm\frac{e}{m}$. For large values of k, we then have

$$E(k) = E_g - \frac{e^{*2}}{\epsilon|\Delta r|} = E_g - \frac{e^2}{m^3 \epsilon k \ell_0^2} \qquad (4.32)$$

from which the gap is estimated. Using the numerical results for the lowest excitation energy obtained for the maximum value of $k\ell_0$ available in the present numerical work, the gap is estimated to be $E_g \sim 0.1$ [4.13] in agreement with the other theoretical results discussed in Chap. 3. Finally, the effect of mixing of higher Landau levels for the excitations has been estimated recently by *Yoshioka* [4.13,15] and found to reduce the energies somewhat.

4.3 Spin Waves

In the earlier sections, we have discussed the role of reversed spins in the ground state and for the quasiparticle-quasihole gap the spin reversed quasiparticles are found to be energetically favorable for low magnetic fields. The collective excitations in the presence of a spin-reversed electron can also be obtained in a similar manner. As discussed earlier in Sect. 3.3, the excited states are classified according to the values of $|S|$ and S_z. In the case of $\nu = \frac{1}{3}$, excited states with $|S| = S_z = \frac{1}{2}N_e$ would correspond to the density wave mode discussed in the preceding section, while for $|S| = S_z = \frac{1}{2}N_e - 1$, the excited state would correspond to the spin-wave excitation.

In Fig. 4.5, we present the numerical results for the low-lying spectrum in the case of one spin-reversed electron in the system at $\nu = \frac{1}{3}$ in the lowest Landau level. Only the three lowest excitation energies are shown in the

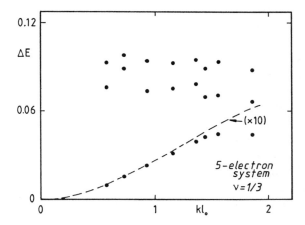

Fig. 4.5. Excitation spectrum for a five-electron system with the spin of one of the electrons reversed relative to the others at $\nu = \frac{1}{3}$. The dashed line is the spin-wave spectrum for $\nu = 1$ (Sect. 5.1)

figure, as a function of $k\ell_0$. Similar results were obtained by *Yoshioka* [4.16]. For small k, the lowest energy mode has a k^2-dependence similar to that obtained by *Kallin* and *Halperin* [4.14] for a fully filled lowest Landau level (Sect. 5.1). For large $k\ell_0$, the lowest energy mode would be the quasiexcitons with a spin-reversed quasiparticle and a spin-polarized quasihole discussed in Sect. 3.3.

A comparison of the density-wave spectrum and the spin-wave spectrum is given in Fig. 4.6 for a five-electron system. The spin-wave spectrum is clearly seen to be lower for all values of $k\ell_0$. The Zeeman energy will provide a constant shift of the spin-wave curve in the upward direction, and would provide a gap at $k = 0$. Considering the magnitude of the gap for small k, it is clear that the spin-wave spectrum would be favored energetically below the momentum at which the minimum of the density wave occurs.

To illustrate the lowest energy neutral excitation spectrum in the presence of a magnetic field, we have presented below some approximate numerical results based on the exact results of Fig. 4.6. We first make a least squares fit through the two sets of points in Fig. 4.6. Strictly speaking, such a fit is not allowed for a five-electron system, since for a finite number of electrons, we only have a finite number of k-values. However, assuming that the five-electron results for the two collective modes are close to those for an the infinite system, qualitatively, such a fit could be considered. The

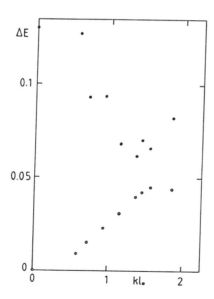

Fig. 4.6. The density-wave spectrum (closed circles) and the spin-wave spectrum (open circles) for a five-electron system at $\nu = \frac{1}{3}$

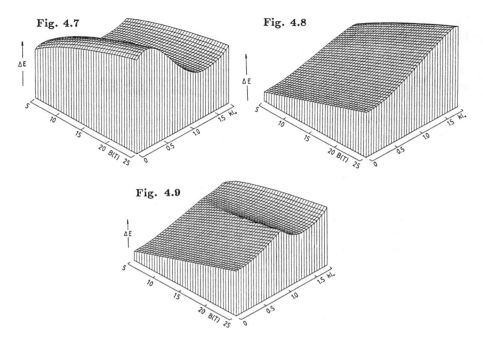

Fig. 4.7. The density-wave mode as a function of $k\ell_0$ and the magnetic field B (in Tesla)

Fig. 4.8. Spin-wave mode as a function of $k\ell_0$ and the magnetic field B (in Tesla)

Fig. 4.9. Lowest energy collective mode as a function of $k\ell_0$ and the magnetic field (in Tesla). For values of $k\ell_0$ below the crossover line, the mode is of spin-wave type, while above the line, the mode is of density-wave type

assumption is drastic, but is reasonable for illustrative purposes. Following the standard procedure of including the finite-thickness correction [see (4.30)], the density-wave mode as a function of magnetic field and $k\ell_0$ is shown in Fig. 4.7.

A similar calculation for the finite-thickness correction is also made for the spin-wave mode, where we have included the Zeeman energy as a function of the magnetic field. The resulting curve is shown in Fig. 4.8, where the linear behavior along the B-axis is due to the dominant Zeeman energy, while the k^2-dependence is clearly noticeable along the $k\ell_0$ axis.

Combining the above two curves and keeping only the *lowest* energy part of the combination, we obtain the result shown in Fig. 4.9. For small $k\ell_0$ the mode is the spin-wave type with the familiar linear behavior with increasing magnetic field, and lower than the density wave mode for all values of magnetic field considered. For $k\ell_0 \sim 1.0$, a crossover line is visible beyond which the density wave mode has the lowest energy, the characteristic minimum of which is discernable in the figure. Note that near the lowest magnetic fields considered here, the spin wave is favorable even for large $k\ell_0$. This should lead to the crossover point observed in Fig. 3.21 for infinitely separated spin-reversed quasiparticle and spin-polarized quasihole pair. However, larger systems are needed to study this region in detail.

As mentioned above, these results are for illustrative purposes only. For a many-electron system, the results will certainly be quantitatively different. However, considering the different momentum dependence of the two modes (density wave and the spin wave), it is conceivable that, such a crossover would exist for a real system. A spin-wave mode has also been obtained in the spherical geometry [4.17]. Finally, considering the valley degeneracy appropriate for Si[110], *Rasolt* and *MacDonald* [4.18] have also studied the the two modes discussed above. Their results will be discussed below.

4.4 Single-Mode Approximation: Magnetorotons

The finite-size calculations discussed above provide quite accurate informations for the collective excitations and the energy gap in the FQHE. However, not much physical insight is gained from these numerical calculations. The theory developed by *Girvin* et al. [4.19], drawing analogies from the Feynman's well-known theory of liquid ^{4}He, however fills the gap. A brief description of the theory is given below. For details, see the paper [4.20] by these authors.

Given the exact ground state ψ, the density-wave excited state at wave vector $\boldsymbol{k}$ is written as [4.21,22],

$$\phi_{\boldsymbol{k}} = N^{-\frac{1}{2}} \rho_{\boldsymbol{k}} \psi \qquad (4.33)$$

with the density operator,

$$\rho_{\boldsymbol{k}} \equiv \sum_{j=1}^{N} e^{i\boldsymbol{k}\cdot\boldsymbol{r}_j} \qquad (4.34)$$

where N is the number of particles. The excitation energy is then given by

$$\Delta(k) = \frac{f(k)}{s(k)}. \qquad (4.35)$$

In (4.35) the denominator is written as

$$f(k) \equiv N^{-1} \langle \psi | \rho_{\boldsymbol{k}}^{\dagger} (\mathcal{H} - E_0) \rho_{\boldsymbol{k}} | \psi \rangle \qquad (4.36)$$

where $\mathcal{H}$ is the Hamiltonian and E_0 is the ground state energy. The function $s(k)$, defined as

$$s(k) \equiv N^{-1} \langle \psi | \rho_k^\dagger \rho_k | \psi \rangle, \qquad (4.37)$$

is the static structure function. The quantity $f(k)$ is in fact the oscillator strength. The final form for the excitation energy is given by the well-known formula,

$$\Delta(k) = \frac{\hbar^2 k^2}{2m s(k)}. \qquad (4.38)$$

The above result can be interpreted as saying that the collective mode energy is the single-particle energy $\hbar^2 k^2/2m$ renormalized by the static structure function representing correlations among the particles.

Defining the dynamic structure factor ($\hbar = 1$) as

$$S(k,\omega) = N^{-1} \sum_n \langle 0|\rho_k^\dagger|n\rangle \delta(\omega - E_n + E_0)\langle n|\rho_k|0\rangle, \qquad (4.39)$$

where the sum is over the complete set of exact eigenstates, and using the relation,

$$s(k) = \int_0^\infty d\omega\, S(k,\omega) \qquad (4.40)$$

we obtain for the oscillator strength,

$$f(k) = \int_0^\infty d\omega\, \omega\, S(k,\omega). \qquad (4.41)$$

These results, when substituted into (4.35) for $\Delta(k)$ show that the Feynman expression for $\Delta(k)$ is in fact the exact first moment of the dynamic structure factor. The quantity $\Delta(k)$ is therefore the average energy for the excitations which are coupled to the ground state through the density. As is well known for liquid ^{4}He there are no low-lying single-particle excitations and the only low-lying excitations are the long-wavelength density oscillations — the *phonons*. The excitation energy curve vanishes linearly, its slope corresponding to the velocity of sound. Near $k = 2\text{Å}^{-1}$, the excitation energy shows a *roton minimum*, which arises due to the peak of the static structure function.

In the case of FQHE, if we insist that the excited state is entirely within the lowest Landau level, the density-wave excited state becomes

$$\phi_k = N^{-\frac{1}{2}} \bar{\rho}_k \psi \qquad (4.42)$$

where $\bar{\rho}_k$ is the *projection* of the density operator onto the subspace of the lowest Landau level. Following the projection technique given by *Girvin* and *Jach* [4.23] (see also Appendix A) in the symmetric gauge we have

$$\bar{\rho}_k = \sum_{j=1}^{N} \exp\left[-ik\frac{\partial}{\partial z_j}\right] \exp\left[-\frac{1}{2}ik^* z_j\right]. \tag{4.43}$$

The projected potential energy is then,

$$\bar{V} = \frac{1}{2} \int \frac{dq}{(2\pi)^2} v(q) \left(\bar{\rho}_q \bar{\rho}_q - \rho e^{-q^2/2}\right) \tag{4.44}$$

where $v(q)$ is the interaction potential ($v(q) = 2\pi/q$ in the present case). Also, the projected oscillator strength is,

$$\bar{f}(k) = N^{-1}\langle 0|\bar{\rho}_k^\dagger[\mathcal{H}, \bar{\rho}_k]|0\rangle \tag{4.45}$$

where $|0\rangle$ is the ground state. Since the kinetic energy is constant, one can write for $\mathcal{H}$ merely the potential energy,

$$\bar{f}(k) = N^{-1}\langle 0|\bar{\rho}_k^\dagger[\bar{V}, \bar{\rho}_k]|0\rangle. \tag{4.46}$$

With the help of the commutation relation for the projected density operators,

$$[\bar{\rho}_k, \bar{\rho}_q] = \left(e^{k^*q/2} - e^{kq^*/2}\right) \bar{\rho}_{k+q}, \tag{4.47}$$

the projected oscillator strength is then

$$\bar{f}(k) = \frac{1}{2}\sum_q v(q) \left(e^{q^*k/2} - e^{qk^*/2}\right) \left[\bar{s}(q)e^{-k^2/2}\left(e^{kq^*/2} - e^{-kq^*/2}\right) \right. \\ \left. + \bar{s}(k+q)\left(e^{k^*q/2} - e^{kq^*/2}\right)\right]. \tag{4.48}$$

In (4.48) the projected static structure factor $\bar{s}(q)$ is defined as

$$\bar{s}(q) = N^{-1}\langle 0|\bar{\rho}_q^\dagger \bar{\rho}_q|0\rangle. \tag{4.49}$$

In terms of the ordinary static structure factor, the projected structure function can be written as

$$\bar{s}(q) = s(q) - \left(1 - e^{-|q|^2/2}\right). \tag{4.50}$$

In the single mode approximation (SMA) (so named because of the assumption that the density-wave alone saturates the full projected oscillator strength sum), the excitation energy is,

$$\Delta(k) = \frac{\bar{f}(k)}{\bar{s}(k)}. \tag{4.51}$$

For small k, $\bar{f}(k)$ vanishes like $|k|^4$ and then for a gap to exist, $\bar{s}(k)$ must vanish as $\sim |k|^4$ [4.20,24]. It is to be noted that the static structure function $s(k)$ is related to the radial distribution function for the ground state $g(r)$ by

$$s(k) = 1 + \rho \int d\boldsymbol{r} \exp(-i\boldsymbol{k} \cdot \boldsymbol{r})\,[g(r) - 1] + (2\pi)^2 \rho \delta^2(k) \qquad (4.52)$$

where ρ is the average density. Expanding (4.52) for small k and using (4.50), one finds that, $\bar{s}(k) \sim |k|^4$, if and only if $M_0 = M_1 = -1$, where

$$M_n \equiv \rho \int d\boldsymbol{r} \left(\tfrac{1}{2}r^2\right)^n [g(r) - 1]. \qquad (4.53)$$

Girvin et al. then expressed the two-body correlation function in terms of the occupation of the single-particle angular momentum eigenstates of (2.19) as:

$$g(r) = \frac{1}{\rho^2} \sum_{\alpha\beta\gamma\delta} \phi_\alpha(0)\phi_\beta(r)\phi_\gamma^*(r)\phi_\delta^*(0) \langle c_\alpha^\dagger c_\beta^\dagger c_\gamma c_\delta \rangle, \qquad (4.54)$$

where $c_\alpha^\dagger$ is the creation operator for the state α. From the conservation of angular momentum and (2.19), one obtains

$$\rho\,[g(r) - 1] = \frac{1}{2\pi\nu} \sum_{m=0}^{\infty} \frac{1}{m!} \left(\tfrac{1}{2}r^2\right)^m e^{\tfrac{1}{2}r^2} \left(\langle n_m n_0 \rangle - \langle n_m \rangle \langle n_0 \rangle - \nu \delta_{m0} \right) \quad (4.55)$$

with $n_m = c_m^\dagger c_m$ being the occupation number for state m. Inserting (4.55) into (4.53), we obtain the result

$$\begin{aligned} M_0 &= \frac{1}{\nu}\left(\langle N n_0 \rangle - \langle N \rangle \langle n_0 \rangle \right) - 1 \\ M_1 &= \frac{1}{\nu}\left[\langle (L+N) n_0 \rangle - \langle L+N \rangle \langle n_0 \rangle \right] - 1 \end{aligned} \qquad (4.56)$$

where $N \equiv \sum_{m=0}^{\infty} n_m$ is the total particle number and $L \equiv \sum_{m=0}^{\infty} m n_m$ is the total angular momentum. As L and N are constants of the motion, their fluctuations vanish and finally one obtains the expected result, $M_0 = M_1 = -1$. The implication is that, in the lowest Landau level, any liquid ground state will have $\bar{s}(k) \sim |k|^4$. If one interprets $\bar{s}(k)$ as the mean-square density fluctuations at wave vector k, the gap condition is then a statement of the incompressibility of the system. The gapless excitations within the SMA can occur only as Goldstone modes with broken translational symmetry.

Using the Laughlin wave function for the ground state, the structure factor and the function $\bar{f}(k)$ can be computed numerically from the above relations. The resulting excitation energy obtained by Girvin et al. is

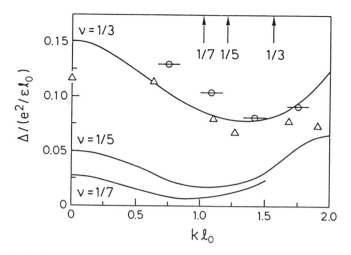

Fig. 4.10. Collective excitation curve in the SMA approximation for $\nu = \frac{1}{3}, \frac{1}{5}$ and $\frac{1}{7}$ filling fractions [4.20]. Arrows at the top indicate magnitude of primitive reciprocal-lattice vector of corresponding Wigner crystal. The points are from the finite-size calculations of [4.1]. Horizontal error bars are the uncertainties in converting angular momentum on the sphere to linear momentum. Triangles are for $N = 6$ periodic boundary condition calculations with a hexagonal unit cell

presented in Fig. 4.10 for $\nu = \frac{1}{3}, \frac{1}{5}$ and $\frac{1}{7}$. The low-lying excitation energy curve reveals several interesting features. The first thing to note is that, unlike ^{4}He, the collective mode has a finite gap at $k = 0$, i.e., the mode is *not* a massless Goldstone mode. It should be pointed out, however, that this gap is *not* due to the charged particles, since the Coulomb force is not sufficiently long-ranged to provide a finite plasma frequency in two dimensions (the plasma frequency goes to zero with a square root dependence on the wave vector). The finite gap originates from the incompressible nature of the electron system at some particular filling fractions.

The collective mode also shows a minimum at a finite k. This minimum is due to the peak in $\bar{s}(k)$ and is thus analogous to the *roton* minimum in liquid ^{4}He. The deepening of the minimum in going from $\nu = \frac{1}{3}$ to $\nu = \frac{1}{7}$ was interpreted by Girvin et al. to be a precursor of the collapse of the gap which would occur at the critical filling ν_c (see Sect. 2.7). The magnitude of the primitive reciprocal-lattice vector of the Wigner crystal, shown by arrows in Fig. 4.10, appears close to the position of the magnetoroton minimum. The collective excitation curve, when compared with the finite system results discussed in earlier sections, shows very good agreement.

Feynman's prediction of the roton energy in ^{4}He was about a factor of two larger than the experimental results. Later *Feynman* and *Cohen*

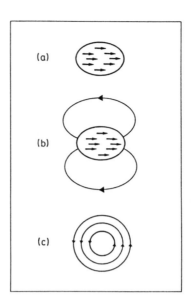

Fig. 4.11 The current distribution in a roton wave packet (schematic): (a) liquid helium with no backflow corrections, (b) helium with backflow corrections, and (c) lowest-Landau-level case [4.20]

[4.22] noted that the roton wave packet made up of the trial wave functions violates the continuity condition:

$$\nabla \cdot \langle \boldsymbol{J} \rangle = 0. \tag{4.57}$$

Let us consider a wave packet,

$$\Phi(r_1,\ldots,r_N) = \int d\boldsymbol{k}\, \xi(k) \rho_{\boldsymbol{k}} \psi(r_1,\ldots,r_N) \tag{4.58}$$

where $\xi(k)$ is a function (e.g. a Gaussian) which is sharply peaked at a wave vector k located in the roton minimum. As the roton group velocity $d\Delta/dk$ vanishes at the roton minimum, the wave packet is quasistationary. Evaluating the current density, one obtains the result shown schematically in Fig. 4.11(a). The current has a fixed direction and is nonzero only in the region localized around the wave packet. The continuity equation is violated becuase the density is approximately time independent for the quasistationary packet. The modified variational wave function of Feynman and Cohen includes the backflow shown in Fig. 4.11(b). The discrepancy with the experimental roton energy was thereby reduced considerably.

In the case of FQHE, the situation is somewhat different. The current density operator is now

$$J(R) = \frac{1}{2m} \sum_j \left[\delta^2(R - r_j) \left(p_j + \frac{e}{c} A(r_j) \right) \right. \\ \left. + \left(p_j + \frac{e}{c} A(r_j) \right) \delta^2(R - r_j) \right]. \quad (4.59)$$

If we take Φ and Ψ to be any two members of the lowest Landau level,

$$\langle \Phi | J(R) | \Psi \rangle = -\frac{1}{2} \nabla \times \langle \Phi | \rho(R) \hat{z} | \Psi \rangle \quad (4.60)$$

where $\rho(R)$ is the density and $\hat{z}$ is the unit vector normal to the plane. From (4.60), it readily follows that

$$\nabla \cdot \langle J(R) \rangle = 0 \quad (4.61)$$

for any state in the lowest Landau level. Therefore, in the FQHE, the backflow condition is automatically satisfied. The current flow for the magnetoroton wave packet is illustrated in Fig. 4.11(c).

Equation (4.61) implies that $\partial \rho / \partial t = 0$ for every state in the lowest Landau level. This is due to the fact that the kinetic energy has been quenched and perturbations can cause the particles to move by (virtual) transitions to higher Landau levels. One should note however that there are, in fact, two different current operators to be considered: The first is the ordinary (instantaneous) current as discussed above. The second one is the slow (time-averaged) $E \times B$ drift of the particles in the magnetic field. If we restrict the Hilbert space to the lowest Landau level, the fast degrees of freedom associated with the cyclotron motion are eliminated, but the slow (drift) one is retained.

For the magnetoroton wave packet, the excess particle density is circularly symmetric. Therefore the (mean) electric field is radial and particle drift is circular [Fig. 4.11(c)]. This is the reason why the continuity condition is automatically satisfied and SMA works so well in the region of the magnetoroton spectrum. Similar results were also obtained by *Saarela* [4.25] using Jastrow theory for inhomogeneous fluids.

In the limit of very large wave vectors ($k\ell_0 \gg 1$), the SMA breaks down. We have seen in earlier sections that in the large k limit, the excitation consists of a quasiparticle-quasihole pair. Therefore for wavelengths much shorter than the interparticle spacing, the density wave is no longer a suitable excitation.

Experimental verification of the dispersion of the magnetoroton modes is difficult in part due to the small number of electrons present within a single layer of two-dimensional electron gas. Recently, *Oji* et al. [4.26] have performed calculations of magnetoroton dispersion in a superlattice composed of electron layers separated by barrier layers. The method is essentially a combination of the SMA for a single layer and a mean-field treatment of interlayer coupling. Their essential result is that, for typical fields and layer spacings, the modes broaden into bands. For sufficiently small layer separation, the energy gap vanishes, and they argue that the gap vanishing is to be associated with an instability toward the Wigner crystal state in each layer. A brief description of their work is given below.

In the lth layer, the induced electron density can be written

$$\delta n_l(q,\omega) = \chi_l(q,\omega)\left[\phi_l^{\text{ext}}(q,\omega) + \sum_{l'\neq l} V_{l'l}(q)\,\delta n_{l'}(q,\omega)\right], \qquad (4.62)$$

with q as the in-plane wave vector, $\phi_l^{\text{ext}}(q,\omega)$ is the frequency dependent external potential in the lth layer, $\chi(q,\omega)$ is the density response function of an isolated layer. The interlayer interaction is given by

$$V_{ll'}(q) = \frac{2\pi e^2}{\epsilon q}\int dz\,dz'\,e^{-q|z-z'|}|\zeta(z-lc)|^2\,|\zeta(z-l'c)|^2 \qquad (4.63)$$

where c is the layer spacing, ϵ is the background dielectric constant, and $\zeta(z)$ is the envelope function. Tunneling of electrons between two layers is forbidden and so there is no interlayer exchange term in (4.62). Multiple scattering between electrons in different layers is expected to be small and is neglected.

In the random phase approximation, (4.62) would be written as

$$\sum_{l'} \left[\Pi^{-1}(q,\omega)\delta_{ll'} - V_{ll'}(q)\right]\delta n_{l'}(q,\omega) = \phi_l^{\text{ext}}(q,\omega) \qquad (4.64)$$

with

$$\chi^{-1}(q,\omega) = \Pi^{-1}(q,\omega) - V(q). \qquad (4.65)$$

Here $\Pi(q,\omega)$ is the proper polarizability of a single layer and $V(q) \equiv V_{ll'}(q)$ is the interlayer interaction.

The collective-mode frequencies are then obtained by requiring

$$\left|\Pi^{-1}(q,\omega)\delta_{ll'} - V_{ll'}(q)\right| = 0. \qquad (4.66)$$

In the SMA approximation, for strong fields and $\omega \ll \omega_c$, we have

$$\chi(q,\omega) = \nu\bar{s}(q)\Delta(q)/\hbar\pi\ell_0^2[\omega^2 - \Delta^2(q)] \qquad (4.67)$$

where $\Delta(q)$ is the single layer magnetoroton dispersion, ν is the filling factor, and $\bar{s}(q)$ is the projected static structure factor given in (4.49,50). For a

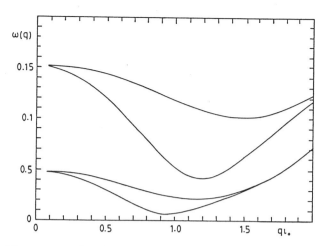

Fig. 4.12. Magnetoroton band dispersion in a superlattice. The higher energy curves are for $\nu = \frac{1}{3}$ and $c = 1.5\ell_0$ while the lower energy curves are for $\nu = \frac{1}{5}$ and $c = 3.5\ell_0$. The maximum and minimum of the band as a function of $q\ell_0$ correspond to $k_z = 0$ and $k_z = \pi/c$ respectively [4.26]

single layer, (4.65–67) give $\omega(q) = \Delta(q)$. For a superlattice ($N_l \to \infty$, N_l being the number of layers)

$$\delta n_l(\mathbf{q}) = \exp(ik_z lc)\, \delta(\mathbf{q}, k_z). \tag{4.68}$$

Assuming zero-thickness of electron layers, (4.65–67) combine to give

$$\omega(q) = \left[\Delta^2(q) + \left(\frac{e^2}{\epsilon \ell_0 \hbar}\right) 2\nu \bar{s}(q)\Delta(q)\left[S(q, k_z) - 1\right]/q\ell_0\right]^{\frac{1}{2}} \tag{4.69}$$

with

$$S(q, k_z) = \frac{\sinh(qa)}{[\cosh(qa) - \cos(k_z a)]}. \tag{4.70}$$

In Fig. 4.12, the extrema of the magnetoroton band are presented as a function of $q\ell_0$ for $c = 1.5\ell_0$, $\nu = \frac{1}{3}$ and for $c = 3.5\ell_0$, $\nu = \frac{1}{5}$. For $qc \gg 1$, $S(q, k_z) \to 1$ and $\omega(q) = \Delta(q)$ and the interlayer coupling vanishes. A similar situation also occurs for $q \to 0$. Near the magnetoroton minimum ($q\ell_0 \sim 1$), $\bar{s}(q)$ reaches a maximum with a substantial effect from interlayer coupling.

For a given value of $q\ell_0$, the magnetoroton modes are stiffest when the layers are oscillating in phase ($k_z = 0$) and softest when the layers are

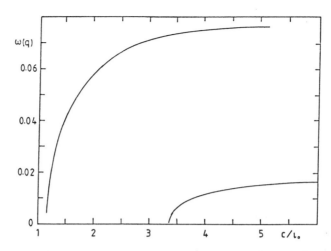

Fig. 4.13. Excitation energy of the superlattice as a function of c/ℓ_0 at $k_z = \pi/c$ for $\nu = \frac{1}{3}$, $q\ell_0 = 1.2$ (upper curve) and $\nu = \frac{1}{5}$, $q\ell_0 = 0.96$ (lower curve) [4.26]

oscillating out of phase. The minimum excitation energy as a function of c/ℓ_0 is plotted for $\nu = \frac{1}{3}$ and $\nu = \frac{1}{5}$ in Fig. 4.13. At $\nu = \frac{1}{3}$, an instability is found to occur when $c/\ell_0 \lesssim 1.5$ and at $\nu = \frac{1}{5}$ for $c/\ell_0 \lesssim 3.35$. *Oji et al.* [4.26] have concluded that the in-plane and k_z wave vectors for the excitations shown in Fig. 4.13 correspond to the fundamental periods of the Wigner crystal states in the superlattice.

As mentioned earlier, within the HF approximation, the ground state of the one-component system is a charge-density wave state with periodicity of a Wigner lattice. For the two-component (spin or valley) system, *Rasolt* and *MacDonald* [4.18] noticed that the exchange interaction occurring only between parallel spins, is expected to favor a fully spin-polarized state for $\nu < 1$ and therefore the same energy per electron as for the one-component system. In Table 4.1, we have reproduced their results for the HF energies of fully polarized and unpolarized CDW states for a series of filling factors. The lowest energy HF state is, in fact, an antiferromagnetic state where the reduction in exchange energy is compensated by the displacement of the positions of up-spin and down-spin charge — thereby reducing the Coulomb energy.

As seen in Table 4.1, for small filling factors, the antiferromagnetic states have energies quite close to those of the CDW state. For larger filling factors, the exchange energy dominates and as a result, the polarized CDW state is strongly preferred at $\nu = 1$.

Rasolt and MacDonald have also calculated the spin-wave dispersion as well as density-wave mode dispersion for various filling factors. Making use

Table 4.1. The energy per electron in the HF approximation for the density wave states for a two-dimensional electron gas at various filling fractions. The unit of energy is $e^2/\epsilon \ell_0$.

ν	Polarized CDW	Unpolarized CDW	Polarized SDW
$\frac{1}{5}$	-0.3220	-0.1636	-0.3195
$\frac{1}{3}$	-0.3885	-0.2204	-0.3823
$\frac{2}{5}$	-0.4123	-0.2491	-0.4013
$\frac{3}{5}$	-0.4838	-0.3162	-0.4332
$\frac{2}{3}$	-0.5076	-0.3354	-0.4393
$\frac{4}{5}$	-0.5505	-0.3526	-0.4798
1	-0.6267	-0.4154	-0.4693

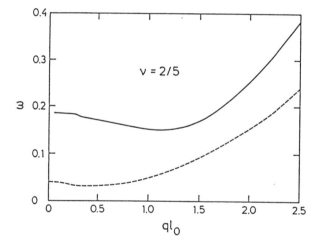

Fig. 4.14 Spin-wave (dashed curve) and density-wave (solid curve) mode dispersion as a function of $q\ell_0$ for the filling factor $\nu = \frac{2}{5}$ [4.18]

of the two-component Laughlin-type state (2.63), their results for the two modes at $\nu = \frac{2}{5}$ are reported in Fig. 4.14.

Su and Wu [4.27] recently reported the excitation spectrum for $\nu = \frac{2}{5}$ calculated in a finite-size system. They found that, in contrast to the $\frac{1}{3}$ case, the collective excitation shows two minima, one at $k\ell_0 \sim 1.6$ and the other one at $k\ell_0 \sim 1.6$. The spectrum calculated from the SMA was found to be inadequate to describe this spectrum.

The specific heat in the case of FQHE ground state has been calculated recently by *Yoshioka* [4.28]. The calculations were done in a spherical geometry. Numerical diagonalization of the Hamiltonian generates the many-particle energy levels E_i of the system. With these energy levels, the partition function and the specific heat are calculated from

$$Z = \sum_i e^{-\beta E_i}$$

$$C = k \sum_i \left[\frac{1}{Z} (\beta E_i)^2 e^{-\beta E_i} \right] - k \left[\sum_i \frac{1}{Z} \beta E_i e^{-\beta E_i} \right]^2 \quad (4.71)$$

with $\beta = 1/k_B T$. The temperature is measured in units of $e^2/\epsilon \ell_0'$, with the modified magnetic length defined by $\ell_0' = (\nu N_s/N_e)^{\frac{1}{2}} \ell_0$ [see (2.56)].

The numerical results for the specific heat evaluated in the spherical geometry for finite-size systems showed a peak around $T = 0.02\, e^2/\epsilon \ell_0'$. The peak is highest at $\nu = \frac{1}{3}$. When ν deviates from $\frac{1}{3}$, the specific heat gets larger at the lower temperature side of the peak and smaller additional peaks appear. The behavior of the specific heat curve has been explained by Yoshioka as due to the presence of a collective mode with a finite energy gap, which gives rise to a Schottky-like peak in the specific heat. There have been some measurements recently on the specific heat of two-dimensional electrons in GaAs-GaAlAs multilayer structures [4.29]. However, in these experiments, the effect of impurities is quite dominant.

To summarize this chapter, from all the theoretical works on finite and infinite systems, the nature of the low-lying collective modes have been well understood. The density-wave mode has been the most thoroughly studied so far. It has a number of characteristic features: For small k, the mode is well separated from the ground state with a large gap. It has a minimum at finite k, which has been found to be analogous to the *roton* minimum present in liquid ^{4}He. The spin-wave spectrum is also quite interesting and needs to be studied further. Unlike the case of the quasiparticle-quasihole gap, no experimental attempt to detect the collective modes has been reported so far. The collective modes are certainly an interesting outcome of the theories initially developed for the stability of the electron system at the experimentally observed filling factors. Experimental verification of these predictions would be an important step in our efforts to understand FQHE.

5. Collective Modes: Inter-Landau Level

In this chapter, we review some of the theoretical work on the magnetoplasmon dispersion. The effect of electron correlations on this mode has been studied in detail by *Kallin* and *Halperin* [5.1,2], for the case of completely filled Landau levels. There has been a considerable amount of experimental work done on the cyclotron resonance in two-dimensional electron systems. The effect of electron correlations on the magnetoplasmon modes might be useful in understanding the anomalous structure in the cyclotron resonance line shape observed in Si-MOSFET's and in GaAs-heterostructures [5.3–7]. For example, experimental observation of cyclotron resonance linewidth broadening and splitting at certain electron densities [5.6] has been attributed to coupling between the cyclotron mode and finite-wavelength magnetoplasmons. A full review of the experimental and theoretical work on this topic will take us beyond the scope of our present review however.[1]

According to Kohn's theorem [5.8], in a system with translational symmetry the electron-electron interactions cannot affect the cyclotron resonance. However, as Kallin and Halperin pointed out, the presence of impurities allows coupling to magnetoplasmon modes at nonzero wave vector, where correlations are important. A direct measurement of the excitation energies is possible in principle [5.9]. The range of wave vectors accessible at present is, however, not wide enough to test the theoretical predictions to be described below. In the following, we present a brief review of the studies of a filled Landau level and compare the results with the case when the lowest Landau level is only *fractionally* filled. In the latter case, the magnetoplasmon mode is strongly influenced by the presence of the intra-Landau level mode discussed in the earlier chapter.

5.1 Filled Landau Level

Here we consider the two-dimensional electron system with no impurity scattering, subjected to a strong perpendicular magnetic field B, and with a density such that an integral number of Landau levels are fully occupied.

---
[1] See the papers by Kallin and Halperin for earlier references.

The elementary neutral excitations near ω_c might then be described as magnetoplasmon modes or *magnetic excitons*, which would correspond to the energy required to promote an electron from an occupied Landau level to unoccupied Landau levels. We consider the strong field limit where the cyclotron energy is large compared to the Coulomb energy. In the absence of electron-electron interactions, the excitation energy would be the kinetic energy difference between the two Landau levels (plus the Zeeman energy difference, if the excitation also involves a spin flip). Inclusion of interaction would shift that energy by an amount of order $e^2/\epsilon\ell_0$.

Neglecting the spin degree of freedom, the wave function for an exciton consisting of an electron in the nth Landau level and a hole in the $(n-1)$th Landau level is calculated (in the Landau gauge) as

$$\psi_{kn}(R, \Delta r) = \frac{1}{2\pi} e^{ik \cdot R} e^{iX\Delta y/\ell_0^2} \phi_n(\Delta r - \ell_0^2 k \times \hat{z}) \tag{5.1}$$

$$\phi_n(r) \equiv \frac{1}{(2n\ell_0^2)^{\frac{1}{2}}} e^{-r^2/4\ell_0^2} \left(\frac{x+iy}{\ell_0}\right) L_{n-1}^1(r^2/2\ell_0^2). \tag{5.2}$$

In (5.1) and (5.2) L_n^α is a Laguerre polynomial, $R \equiv \frac{1}{2}(r_1 + r_2) = (X, Y)$ and $\Delta r \equiv r_1 - r_2 = (\Delta x, \Delta y)$ with r_1 and r_2 being the electron and hole positions respectively. The vector k plays the role of the total momentum of the particles.

The exciton wave function (5.1) is a direct product (except for a gauge independent phase factor) of a plane wave in the center of mass coordinates R and a function of the relative coordinates $\phi_n(r)(\Delta r - \ell_0^2 k \times \hat{z})$, whose magnitude is spherically symmetric about the point $\Delta r = \ell_0^2 k \times \hat{z}$. The dipole moment of the exciton can then be defined as

$$e\langle\psi_{kn}|\Delta r|\psi_{kn}\rangle = e\ell_0^2 k \times \hat{z} \tag{5.3}$$

which is perpendicular to k, proportional to k and independent of n. Classically this is what one expects as the two particles of opposite charge in a magnetic field move parallel to one another (in contrast to the *electrons* in a magnetic field, which orbit one another) with a constant linear velocity perpendicular to their separation. The exciton momentum increases with increasing separation between the particles, even though the velocity decreases.

With the exciton wave function (5.1), Kallin and Halperin have calculated the exciton dispersion relation. The result is *exact* to lowest order in $(e^2/\epsilon\ell_0/\hbar\omega_c)$, where the contributions from the particle-hole ring and ladder diagrams have been included self-consistently. In the case where one spin state is occupied in the lowest Landau level, the exciton energy is given explicitly as

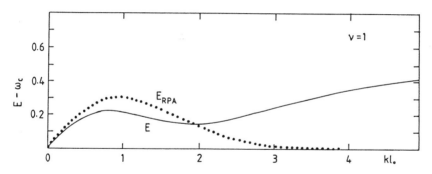

Fig. 5.1. Exciton dispersion curve near ω_c for $\nu = 1$ (spin polarized). The dotted curve is obtained within the random phase approximation (RPA) [5.1]

$$E(k) = \omega_c + \frac{e^2}{2\epsilon\ell_0}\left[(\pi/2)^{\frac{1}{2}}\left\{1 - e^{-k^2\ell_0^2/4}\left(\left(1 + \frac{1}{2}k^2\ell_0^2\right)\right.\right.\right. \\ \left.\left.\left.\times I_0\left(\frac{1}{4}k^2\ell_0^2\right) - \frac{1}{2}k^2\ell_0^2\, I_1\left(\frac{1}{4}k^2\ell_0^2\right)\right)\right\} + \nu k\ell_0 e^{-k^2\ell_0^2/2}\right] \quad (5.4)$$

where I_n is a modified Bessel function. The resulting curve is plotted in Fig. 5.1. The exciton dispersion curve has a maximum at $k\ell_0 \sim 0.9$ and a minimum at $k\ell_0 \sim 2$. The energy shift goes to zero as $k\ell_0 \to 0$, in accordance with Kohn's theorem. For $k\ell_0 \ll 1$, the spectrum is seen to increase *linearly* from the origin as $\Delta E(k) = E(k) - \omega_c = \frac{1}{2}k\ell_0$. Kallin and Halperin also calculated the dispersion within the random phase approximation (RPA), which is seen to be clearly inadequate to describe the mode for large $k\ell_0$ (Fig. 5.1).

MacDonald [5.10] did a similar calculation where he also included higher interaction strength within the HF approximation. The net effect was found to shift the dispersion curve down from that of the leading order result.

In the case of one spin component occupied in the lowest Landau level, one can also calculate the spin wave spectrum. The dispersion relation for this case was obtained by Kallin and Halperin to be,

$$E(k) - |g\mu_B B| = \frac{e^2}{\epsilon\ell_0}\left(\frac{1}{2}\pi\right)^{\frac{1}{2}}\left[1 - e^{-k^2\ell_0^2/4}I_0(k^2\ell_0^2/4)\right], \quad (5.5)$$

where the second term on the left hand side is the usual Zeeman energy. The result is plotted in Fig. 5.2. The energy shift approaches zero in the limit $k \to 0$. Similar results were also derived by *Bychkov* et al. [5.11].

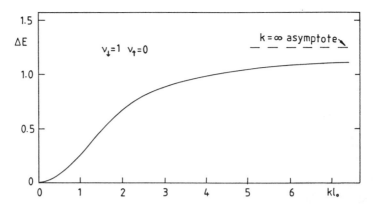

Fig. 5.2. Spin wave spectrum for $\nu = 1$. The Zeeman energy is not included [5.1]

5.2 Fractional Filling: Single Mode Approximation

The single mode approximation has been discussed in Sect. 4.4 in the context of intra-Landau level collective excitations of fractional quantum Hall states. This approach has been extended by *MacDonald* et al. [5.12] to the present problem of inter-Landau level excitations. Their work is primarily concerned with the magnetoplasmon excitations in the extreme quantum limit where the ground state $|\Psi_0\rangle$ is strictly within the lowest Landau level. In the following we shall present a brief description of their approach.

Let us consider the symmetric gauge where the single particle eigenfunctions are given by (see Appendix A)

$$\Psi_{nm}(z_i, z_i^*) = \frac{1}{(2\pi)^{\frac{1}{2}}} \frac{(a_i^\dagger)^n (b_i^\dagger)^m}{(n!m!)^{\frac{1}{2}}} e^{\frac{1}{4} z_i z_i^*}. \tag{5.6}$$

They are eigenfunctions of the kinetic energy operator

$$\begin{aligned}\mathcal{K}_i \Psi_{nm}(z_i, z_i^*) &= \hbar\omega_c \left(a_i^\dagger a_i + \tfrac{1}{2}\right) \Psi_{nm}(z_i, z_i^*) \\ &= \hbar\omega_c \left(n + \tfrac{1}{2}\right) \Psi_{nm}(z_i, z_i^*).\end{aligned} \tag{5.7}$$

Here, (as discussed in Appendix A) $a_i^\dagger$ and $b_i^\dagger$ are inter-Landau level and intra-Landau level ladder operators respectively. The intra-Landau level collective excitations were created by the projected density operator $\bar{\rho}$. In the present case, however, we must also include other levels. To proceed in an analogous way we therefore decompose the full density operator ρ_k into a sum over Landau levels

$$\rho_k = \sum_{n'n} \rho_k^{n'n} \tag{5.8}$$

where the operator

$$\rho_k^{n'n} = \sum_i A_i^{n'n}(k)B_i(k) \qquad (5.9)$$

transfers particles from level n to level n'. The operator $B_i(k)$ is an intra-Landau level operator and is defined in Appendix C. The inter-Landau level operator $A_i^{n'n}$ (also given in Appendix C) which can further be written in the form

$$\begin{aligned}A_i^{n'n}(k) &= \exp\left(-\frac{i}{\sqrt{2}}k a_i^\dagger\right) \exp\left(-\frac{i}{\sqrt{2}}k^* a_i\right) \\ &= |n'\rangle_{ii}\langle n|G^{n'n}(k)\end{aligned} \qquad (5.10)$$

is responsible for the transfer of the particle i from the level n to the level n'. The notation $|n\rangle_i$ describes the state where the particle i is promoted to the level n while all the other particles remain in their original level. The coefficient in the last form is given by

$$G^{n'n}(k) = \sqrt{\frac{n!}{n'!}} \left(-\frac{ik}{\sqrt{2}}\right)^{n'-n} L_n^{n'-n}(\tfrac{1}{2}k^2). \qquad (5.11)$$

The wave functions of the excited states of momentum k are expanded in terms of the states $\rho_k^{n'n}|\Psi_0\rangle$

$$|\Psi_k\rangle = \sum_{n',n} \alpha_{n'n}(k)\rho_k^{n'n}|\Psi_0\rangle, \qquad (5.12)$$

thereby generalizing the SMA for the intra-Landau level excitations. The diagonalization of the Hamiltonian in the space of these functions leads to the secular equation

$$\sum_{n',n}\left[E_{\text{pl}}(m',m;n',n:k) - \Delta(k)S(m',m;n',n:k)\right]\alpha_{n'n}(k) = 0. \qquad (5.13)$$

Here E_{pl} is the matrix element of the Hamiltonian, minus the ground state energy, between the states in the sum (5.12)

$$E_{\text{pl}}(m',m;n',n:k) = \langle\Psi_0|\rho_{-k}^{mm'}[\mathcal{H},\rho_k^{n'n}]|\Psi_0\rangle. \qquad (5.14)$$

The scalar products of these states are denoted by

$$S(m',m;n',n:k) = \langle\Psi_0|\rho_{-k}^{m'm}\rho_k^{n'n}|\Psi_0\rangle \qquad (5.15)$$

and are straightforward generalizations of the projected structure function $\bar{s}$ defined by (4.49). The part containing the kinetic energy operator in the commutator appearing in (5.14) is easily evaluated to be

$$[\mathcal{K},\rho_k^{n'n}] = \hbar\omega_c(n'-n)\rho_k^{n'n}. \qquad (5.16)$$

Substituting this and the Fourier transform for the potential energy into (5.14), one can rewrite E_{pl} as

$$E_{\text{pl}}(m',m;n',n:k) = \hbar\omega_c(n'-n)S(m',m;n',n:k)$$
$$+ \frac{1}{2}\int \frac{d^2q}{(2\pi)^2} V(q)\langle\Psi_0|\rho_k^{mm'}[\rho_{-q}\rho_q,\rho_k^{n'n}]|\Psi_0\rangle. \quad (5.17)$$

As an application of the formalism, two limiting cases are considered. The first one is the extreme quantum limit case where all excitations are restricted to the lowest Landau level. The secular equation (5.13) then reduces to

$$\Delta(k) = \frac{E_{\text{pl}}(0,0;0,0:k)}{S(0,0;0,0:k)} \quad (5.18)$$

which is clearly equivalent with the expression (4.51) for intra-Landau level excitations. The second case corresponds to the coherent promotion of an electron from the lowest level to nth Landau level. In the extreme quantum limit when there is no Landau level mixing, the secular equation again reduces to

$$\Delta(k) = \frac{E_{\text{pl}}(n,0;n,0:k)}{S(n,0;n,0:k)}. \quad (5.19)$$

In this special case the function S takes a particularly simple form

$$S(n,0;n,0:k) = \langle\Psi_0|\rho_k^{0n}\rho_k^{n0}|\Psi_0\rangle = N_e\left|G^{n0}(k)\right|^2 e^{-\frac{|k|^2}{2}} \quad (5.20)$$

as can be seen by applying the algebra obeyed by the operators $B_i(k)$

$$B_i(k_1)B_i(k_2) = e^{\frac{1}{2}k_1^*k_2} B_i(k_1+k_2) \quad (5.21)$$

and using the fact that

$$_i\langle n| \cdot |n\rangle_j = \delta_{ij}. \quad (5.22)$$

Let us now introduce the notation

$$h(q) = s(q) - 1$$
$$\widetilde{h}(q) = e^{\frac{1}{2}|q|^2} h(q) \quad (5.23)$$

and make use of the identities

$$\sum_{r\neq s}\langle\Psi_0|B_r(-k)B_s(k)|\Psi_0\rangle = N_e[h(k) + 2\pi\nu\delta^2(k)]$$
$$\sum_l G^{nl}(k_1)G^{lm}(k_2) = e^{-\frac{1}{2}k_1^*k_2} G^{nm}(k_1+k_2) \quad (5.24)$$

we obtain the final form for E_{pl} (5.17):

$$E_{\text{pl}}(n,0;n,0:k) = S(n,0;n,0:k)\left(\int \frac{d^2q}{(2\pi)^2} e^{-\frac{1}{2}|q|^2} V(q)\right.$$
$$\times \left\{\tilde{h}(q)\left[G^{nn}(q)e^{\frac{1}{2}(k^*q - kq^*)} - 1\right] + \tilde{h}(k-q)|G^{n0}(q)|^2\right\}$$
$$\left. + n\hbar\omega_c + \frac{\nu e^2}{k}|G^{n0}(k)|^2 e^{-\frac{1}{2}|k|^2}\right)$$
(5.25)

where the factor inside the big parenthesis is the SMA result for $\Delta(k)$. By considering the projected f sum rule it can be shown that for the strong field limit this is an exact result provided that a single mode exhausts all of the projected oscillator strength.

The SMA result (5.25) can be compared for example with the summation of the self-consistent Hartree-Fock ladder diagrams performed by *Oji* and *MacDonald* [5.13]

$$\Delta(k) = n\hbar\omega_c + H^n(k) + X^n(k). \tag{5.26}$$

The Hartree term

$$H^n(k) = \frac{\nu e^2}{k}|G^{n0}(k)|^2 e^{-\frac{1}{2}|k|^2} \tag{5.27}$$

is equal to the last term in the SMA. The random phase approximation, which neglects all of the many-body corrections, is obtained by including the cyclotron energy and the Hartree term. It thus corresponds to the last two terms in (5.23). Many-body corrections due to the exchange effects are taken into account by the Fock term

$$X^n(k) = -\nu \int \frac{d^2q}{(2\pi)^2} e^{-\frac{1}{2}|q|^2} V(q) \left\{G^{nn}(q)e^{\frac{1}{2}(k^*q - kq^*)} - 1 + |G^{n0}(q)|^2\right\}. \tag{5.28}$$

We can see that the SMA reduces to the Hartree-Fock expression when the function $\tilde{h}(q)$ is replaced by its uncorrelated value, $-\nu$. The SMA is then expected to be an excellent approximation since its takes into account the many-body effects including the electron-electron correlations via the function $\tilde{h}(q)$ which is plotted in Fig. 5.3 for various values of the filling fraction ν. The calculations of $\tilde{h}(q)$ are based on Laughlin's ground state wave functions and therefore introduce an approximation to the otherwise exact result (5.25). Nevertheless these wave functions are known to be extremely accurate. The excitation energies for the special case where an electron is promoted from the lowest level to the next level are presented in Fig. 5.4 for the values $\nu = 1, \frac{1}{3}, \frac{1}{5}$ and $\frac{1}{7}$. Compared with the Hartree-Fock results there is a tendency to make the magnetoplasmon mode stiffer which is consistent with the observed narrowness of the cyclotron resonance line found by *Wilson* et al. [5.3,4]. Later measurements however have

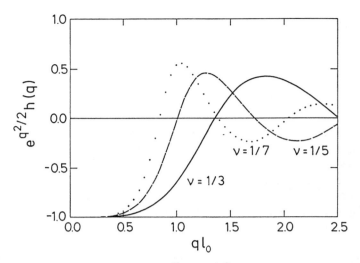

Fig. 5.3. Projected correlation function $\widetilde{h}(q) = e^{|q|^2} h(q)$ for the fractional Hall states at $\nu = \frac{1}{3}$ (solid line), $\nu = \frac{1}{5}$ (dashed line) and $\nu = \frac{1}{7}$ (dotted line) [5.12]

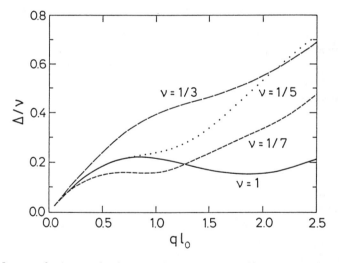

Fig. 5.4. Magnetoplasmon excitation energies measured from $\hbar\omega_c$ (in units of $\nu e^2/\ell_0$) for $\nu = 1$ (solid line), $\nu = \frac{1}{3}$ (long-dashed line), $\nu = \frac{1}{5}$ (dotted line) and $\nu = \frac{1}{7}$ (short-dashed line) [5.12]

revealed that there is a broadening or splitting of this line at certain values of the perpendicular magnetic field [5.6,7]. It is expected that the softening of the magnetoplasmon modes at some higher values of the wave vector is responsible for this observed behavior. One way to improve the SMA scheme would be to couple the modes to the intra-Landau level modes which would cause a broadening of the magnetoplasmon excitations.

In addition to the coherent excitations described by the SMA there are also incoherent excitations. They are somewhat analogous to the particle-hole excitations in an ordinary Fermi liquid in the sense that the particle promoted to a higher Landau level leaves a hole behind it. In this case, however, the momentum of the hole is exactly opposite to the momentum of the particle. Therefore the total momentum of the excitation is determined by the state from which the electron was removed. In Appendix C we show that the energy of this incoherent excitation mode is an exact repetition, when shifted by $\hbar\omega_c$, of the intra-Landau level mode. Comparing the dispersion curves presented in Figs. 4.10 and 5.4 it can be seen that the incoherent mode is energetically more favorable at the values of the wave vector $q\ell_0 \gtrsim 0.7$. The incoherent mode is still well above the $\hbar\omega_c$-level and hence cannot be accounted for the splitting of the cyclotron resonance line. In the presence of impurities these two modes might couple, which would bring one of them near the cyclotron energy.

5.3 Fractional Filling: Finite-Size Studies

Let us now consider the inter-Landau level excitations in a system with a finite number of electrons under exactly the same geometrical conditions as discussed in Sect. 4.2. The symmetry analysis described there can straightforwardly be extended to cover the system where electrons are allowed to occupy arbitrary Landau levels. However, as more particles are distributed among Landau levels, this extra degree of freedom will extremely rapidly increase the number of state vectors needed to describe the system. Therefore, in practice, one can study only those systems where only one of the electrons is promoted from the lowest level and the effects of the Landau level mixing are ignored.

As a special case we consider the system of N_e electrons where one electron is elevated to Landau level 1 with the remaining $N_e - 1$ electrons in the lowest level. As before, we work in the occupation number space. The N_e-electron state is described by the state vector (see Appendix C for more details)

$$|L; k\rangle = |j_1 0, j_2 0, \ldots, j_k 1, \ldots, j_{N_e} 0\rangle \qquad (5.29)$$

which can be interpreted as the state where the electron k lies in the level 1. The complete set of states (5.29) can now be divided into equivalence classes, just like we did in Sect. 4.2, but with the equivalence relation (4.25) replaced by

$$|j'_1 K'_1, j'_2 K'_2, \ldots, j'_{N_e} K'_{N_e}\rangle$$
$$= |j_1 - qk\, K_1, j_2 - qk\, K_2, \ldots, j_{N_e} - qk\, K_{N_e}\rangle \quad (5.30)$$

in order to incorporate the Landau level labelling. From the state (5.29) a momentum eigenstate is constructed according to formula (4.27)

$$|(s,t); \mathcal{L}\rangle$$
$$= \frac{1}{\sqrt{|\mathcal{L}|}} \sum_{k=0}^{|\mathcal{L}|-1} \exp\left\{\frac{2\pi i s}{N} k\right\} |j_1 - qk\, K_1, j_2 - qk\, K_2, \ldots, j_{N_e} - qk\, K_{N_e}\rangle. \quad (5.31)$$

It should be noted that compared with the intra-Landau level system the number of equivalence classes $\mathcal{L}$, and accordingly the number of relative momentum eigenstates $|(s,t); \mathcal{L}\rangle$, are now more than N_e times greater. Firstly, any one of the N_e electrons can be promoted to the next level, with each one of the choices creating a new class. Secondly, it is now possible to doubly occupy a given single-particle state labelled by the quantum number j.

The Hamiltonian, which is diagonalized in the space spanned by the states (5.31), is given in (4.28-31). The kinetic energy part of the Hamiltonian (4.28) is of course already diagonal contributing the constant $\hbar\omega_c$ to the total energy and could therefore be omitted. On the other hand it represents the *total* kinetic energy and from the treatment presented in Sect. 4.2 we know that the physically relevant quantities are *relative*. The kinetic energy $\mathcal{K}$ is therefore split into two parts, $\mathcal{K} = \mathcal{K}^{\text{CM}} + \mathcal{K}^{\text{R}}$. The physical state must be an eigenstate of the relative kinetic energy operator

$$\mathcal{K}^{\text{R}} = \frac{1}{2mN_e} \sum_{l<k} (\Pi_l - \Pi_k)^2 \quad (5.32)$$

or, since the states are by construction eigenstates of $\mathcal{K}$, an eigenstate of the center of mass operator $\mathcal{K}^{\text{CM}}$.

A numerical calculation based on the formalism presented above was carried out by *Pietiläinen* and *Chakraborty* [5.14]. The results are presented in Fig. 5.5 for the filling fraction $\nu = \frac{1}{3}$, where only the lowest excitation energies are presented. The most striking feature of these results is that they exactly repeat the roton spectrum in the lowest Landau level. In Appendix C, we have presented a proof (see also [5.15]) that among the inter-Landau level excitations there are points which exactly reproduce the intra-Landau excitations when shifted by $\hbar\omega_c$. The numerical studies show that these repeated points are energetically the lowest modes. This is clearly in contradiction with the SMA results. One should note however that, in practice, it is not possible in a finite system to study the behavior of the dispersion curve at small values of the wave vector where the SMA is supposed to be at its best.

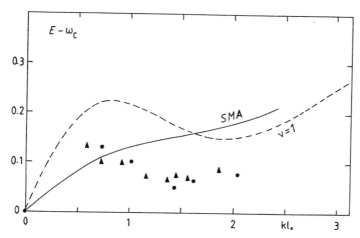

Fig. 5.5. Shift of the lowest energy mode from $\hbar\omega_c$ obtained for $\frac{1}{3}$ filling of the lowest Landau level (Spin polarized). Only the lowest excitation energies are shown for each $k\ell_0$, for four (●) and five-electron (▲) systems. The dashed curve is from [5.1]. The SMA curve is for $\nu = \frac{1}{3}$ from [5.12]

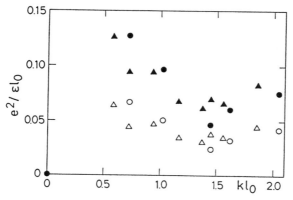

Fig. 5.6. The finite electron system result of Fig.5.5 as function of the dimensionless thickness parameter $\beta = (b\ell_0)^{-1}$. The filled and empty points are for $\beta = 0$ and $\beta = 1$ respectively

The qualitative features of the finite system results are in agreement with the spectrum where the lower Landau level is fully occupied [5.1,2]. A minimum at a finite wave vector is conspicuously present. The energy shift is smaller than in the $\nu = 1$ case, which is again expected on general grounds [5.1]. The minimum at the finite wave vector can soften and move into resonance with the cyclotron mode, if we introduce the standard procedure

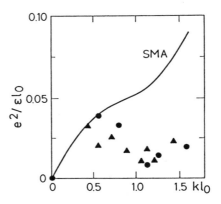

Fig. 5.7. Same as in Fig. 5.5 but now for $\nu = \frac{1}{5}$ for four-(●) and five-electron (▲) systems and the SMA result is from [5.12]

of finite-thickness corrections. In Fig. 5.6, we have plotted the spectrum for two values of the dimensionless thickness parameter $\beta = (b\ell_0)^{-1}$ (see Sect. 2.6). The magnetic fields employed so far in the QHE experiments should lie between those values of β.

In Fig. 5.7, we have presented the four- and five-electron results for $\nu = \frac{1}{5}$. This filling factor is interesting since in this case, the SMA result shows a shallow minimum developing around $k\ell_0 \sim 1$, in contrast to the almost monotonic increase of the $\nu = \frac{1}{3}$ spectrum in the SMA (see Fig. 5.5). The change in shape of the spectrum for $\frac{1}{5}$ filling, as explained by *MacDonald* et al. [5.12], is directly related to the structure factor and hence to the correlations which are ν-dependent. Qualitatively, no difference between $\frac{1}{3}$ and $\frac{1}{5}$ fillings in the finite-size calculations is seen in Fig. 5.7.

As pointed out in the previous section and proved in Appendix C, the repetition is not a property of a finite-size system but manifests itself also in a many particle system. From the above studies of the magnetoplasmon spectrum with fractional filling of the lower Landau level the following picture therefore emerges: At small values of the wave vector, $k\ell_0 \lesssim 0.7$, judging from Figs. 4.10, 5.4 and 5.5, the coherent magnetoplasmon mode (5.12) is energetically preferable while at larger values the incoherent mode takes over thus exhibiting the repetition of the magnetoroton mode.

6. Further Topics

In this chapter, we present a mixed bag of results which are interesting and of importance for our understanding of FQHE. Much of the work described below needs further development and our aim in this chapter is simply to emphasize the important aspects of the presently available results.

6.1 Effect of Impurities

The effect of impurities on the Laughlin ground state has been investigated by *Pokrovskii* and *Talapov* [6.1,2]. They considered a delta-function form for the impurity interaction:

$$V_i(r) = V_i\,\delta(r). \tag{6.1}$$

Recall that for $\nu = \frac{1}{m}$, Laughlin's state is nondegenerate. For $\nu < \frac{1}{m}$, the ground state is multiply degenerate, and the wave function is written

$$\psi = \psi_m\, Q(z_1,\dots,z_{N_e}) \tag{6.2}$$

where Q is a symmetric polynomial of order $s \leq N_s - m(N_e - 1)$ in each variable [see (2.25)]. The impurities lift the degeneracy. Since V_i is positive, the wave function should vanish at points where the impurities are located. We then have

$$\psi = \prod_{\substack{i\leq j\leq N_e \\ 1\leq k\leq N_i}} (z_j - \eta_k)\, \psi_m(z_1,\dots,z_{N_e}) \tag{6.3}$$

where η_k is a complex coordinate of the kth impurity, and N_i is the number of impurities. Such a state can be realized only if $N_i \leq s$. Comparing this state with (3.2), one can interpret the state (6.3) as the one in which a Laughlin quasihole of charge $+\frac{|e|}{3}$ is trapped at each impurity.

For $N_i > s$, we cannot distribute the quasiholes among all the impurities. The ground state in this case would be written as

$$\psi = \prod_{\substack{1\leq j\leq N_e \\ 1\leq k\leq s}} (z_j - \eta_k)\, \psi_m(z_1,\dots,z_{N_e}) \tag{6.4}$$

where the s impurities are chosen from N_i impurities such that the energy is minimum. The ground state energy is

$$E_0 = V_i \sum_{s < k \leq N_i} \rho(\eta_k) \tag{6.5}$$

where $\rho(\eta)$ is the electron density at η. Far from the holes one must have $\rho(\eta) \approx N_e/N_s$. The above authors have also studied the case of a strong impurity potential and its effect on the Hall steps [6.2].

In a detailed study of the effect of a charged impurity in the FQHE, Zhang et al. [6.3] performed finite-size system calculations in different geometries. They considered the electron-impurity interaction term in the Hamiltonian

$$\mathcal{H}_i = -\sum_j \frac{Ze^2}{\epsilon|\boldsymbol{R}_i - \boldsymbol{r}_j|} + E_{i-b} \tag{6.6}$$

where Ze is the impurity charge, ϵ is as usual, the background dielectric constant, $\boldsymbol{R}_i$ is the impurity position, $\boldsymbol{r}_j$ is the position of the jth electron, and E_{i-b} is the constant impurity-background interaction energy. Only the spin-polarized state was considered. The electron Hamiltonian with the additional term $\mathcal{H}_i$ is then diagonalized numerically for few-electron systems ($N_e = 3$ to 6).

In the spherical geometry as we recall, the ground state occurs for $L = 0$, where L is the total angular momentum. The presence of an impurity breaks the spherical symmetry. However, the azimuthal symmetry is still preserved and the states could be classified according to L_z. The ground state is at $L_z = 0$ and the excited states which are degenerate, split, because of level

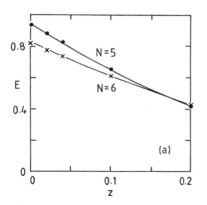 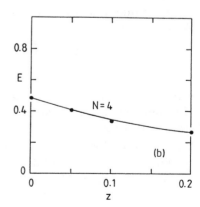

Fig. 6.1. Excitation energy gap at $\nu = \frac{1}{3}$ as a function of the impurity strength Z for (a) spherical, and (b) rectangular geometry for a finite number of electrons [6.3]

mixing by the impurity potential. In Fig. 6.1(a), we present the results of Zhang et al. for the excitation gap in the spherical geometry (defined as the difference between the ground state and the lowest excited state), as a function of the impurity strength Z for $N_e=5$ and 6. A significant reduction of the gap is noticeable in the result. The screening charge density for different impurity strengths is plotted in Fig. 6.2(a) in the spherical

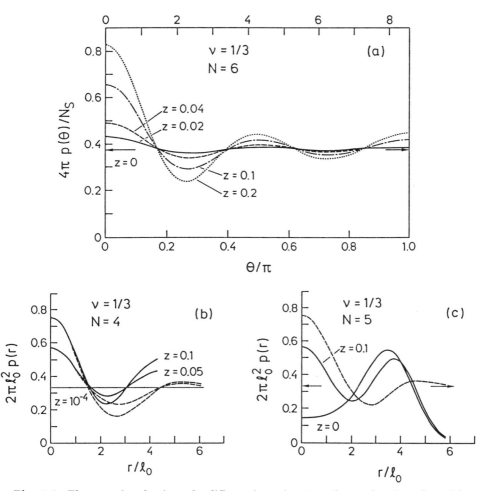

Fig. 6.2. The screening density ρ for different impurity strength as a function of spatial separation in (a) spherical [$\rho(\theta)$ is the density per unit solid angle], (b) rectangular, and (c) disk geometries. In the rectangular geometry, the rotational symmetry is absent; the solid curves show the charge density in the (0,1) direction and the dashed curve is that in the (1,1) direction. In (c) the dashed curve gives the screening charge density. The arrows indicate the average normalized charge density in the disk at $\nu = \frac{1}{3}$ [6.3]

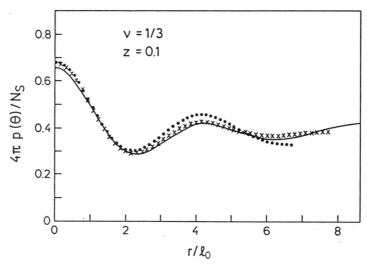

Fig. 6.3. The screening charge in the spherical geometry for a fixed charge $Z = 0.1$ as a function of the interparticle separation for different system sizes: $N_e = 6$ (solid curve), $N_e = 5$ (crosses), and $N_e = 4$ (dots) [6.3]

geometry. The screening charge accumulates at the impurity and oscillates away from the impurity with a characteristic length scale of ℓ_0. For fixed Z, the charge density is plotted in Fig. 6.3 for N_e=4, 5 and 6. The screening is quite independent of the system size.

Zhang et al. performed similar calculations for other geometries. In a periodic rectangular geometry, the impurity-free ground state is triply degenerate at $\nu = \frac{1}{3}$. The degeneracy is lifted by the impurity potential which mixes the momentum eigenstates. Again defining the excitation gap as the energy difference between the *lowest* ground state and the *lowest* excited state levels, the gap is plotted in Fig. 6.1(b) for N_e=4. The screening behavior for the rectangular geometry is presented in Fig. 6.2(b). These results are qualitatively similar to those for the spherical geometry.

In the disk geometry, there is no downward cusp in the ground state energy, which might be related to the open boundary condition (edge effect). The electronic charge density in the disk geometry both with and without the impurity (placed at the center of the disk), is shown in Fig. 6.2(c). For the impurity-free case, the charge density is *nonuniform*, in contrast to the other two geometries considered, and is due to the finite-size effect. In the presence of the impurity, the screening behavior is very similar to that of the other two geometries.

The screening oscillation is not related to the Friedel oscillation, which arises due to the existence of a sharp Fermi surface, but as these authors pointed out, is a consequence of the incompressiblity of the system. Similar calculations have been reported by *Rezayi* and *Haldane* [6.4] for a spherical geometry.

Rezayi and Haldane studied the effect of a short-range or delta-function impurity potential. Any potential, whose range is much less than the magnetic length could effectively be considered as a delta function with binding energy,

$$g = \frac{1}{\pi \ell_0^2} \int d\mathbf{r}\, V(\mathbf{r})\, e^{-(r/\ell_0)^2}$$

for particles in the lowest Landau level. In fact, the real-space form of the linear response function in the ground state is defined by the charge density response to a weak delta function impurity potential. For a six-electron system in a spherical geometry, it was found that, at the position of the impurity, the charge density increases from zero to the maximum value as the impurity strength varies from $-\infty$ to $+\infty$. In the neighborhood of the impurity, there is a local oscillatory polarization of the charge density. The period of this oscillation is governed by the linear response of the Laughlin state and remains essentially unchanged even well outside the linear response regime.

The above authors also studied the effects of Coulomb impurity potentials for the six-electron system at $\nu = \frac{1}{3}$. The calculations are similar to those by *Zhang* et al. [6.3], described above. The only difference is that, in the calculations of *Rezayi* and *Haldane* [6.4], the charge of the impurity was varied from $+0.5e$ to $-0.5e$. They noticed that, in contrast to short-range impurities, the incompressible state was unstable if the charge of the impurity exceeded certain critical value. For the six-electron system the critical charges were found to be $+0.38e$ and $-0.30e$.

An interesting upshot of the above studies is the result for the linear response [6.5]. The dynamic structure function in the SMA is given by

$$S(q,\omega) = \bar{s}(q)\, \delta\!\big(\omega - \Delta(q)\big). \tag{6.7}$$

The static susceptibility to an external perturbation is defined as

$$\begin{aligned}\chi(q) &= -2 \int_0^\infty \frac{d\omega}{\omega}\, S(q,\omega) \\ &= -\frac{2\bar{s}(q)}{\Delta(q)}.\end{aligned} \tag{6.8}$$

The quantity $\alpha(q) \equiv \bar{s}(q)/\Delta(q)$ is plotted in Fig. 6.4.

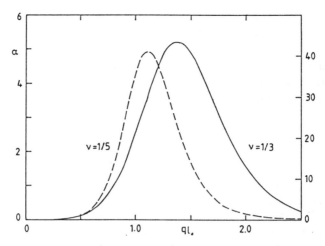

Fig. 6.4. Susceptibility $\alpha = -\frac{1}{2}\chi(q)$ for $\nu = \frac{1}{3}$ (scale on left) and $\frac{1}{5}$ (scale on right) [6.5]

Within linear response theory we have

$$\langle \delta\bar{\rho}_q \rangle = \rho v(q)\chi(q) \tag{6.9}$$

where $v(q) = 2\pi Z e^2/\epsilon q$ is the Fourier transformed Coulomb interaction and the mean density $\rho = e\nu/2\pi\ell_0^2$. The space charge distribution is

$$\frac{\langle \rho(r) \rangle}{\rho} = 1 + Z \int_0^\infty dq\, \chi(q) J_0(qr) \tag{6.10}$$

and is shown in Fig. 6.5. As $\chi(q)$ is sharply peaked at the roton wave vector, the spatial distribution of the charge is oscillatory. This is the simple physical explanation provided by *Girvin* et al. [6.5], for the charge oscillation observed by *Zhang* et al. [6.3], discussed above (Fig. 6.3). The impurity relaxation energy in linear response is given by

$$\Delta E = \frac{1}{2} \int \frac{dq}{(2\pi)} v(q) \langle \delta\bar{\rho}_q \rangle. \tag{6.11}$$

For $\nu = \frac{1}{3}$, Girvin et al. obtained the value $\Delta E = -1.15 Z^2$. For a five-electron system, the result is $\Delta E = -1.2 Z^2$ [6.3].

Tao and *Haldane* [6.6] studied the impurity effect on FQHE considering the case of finite systems in a periodic rectangular geometry. They found that weak impurities make the ground state quasidegenerate if there is an energy gap. The Hall conductance is directly related to the first chern

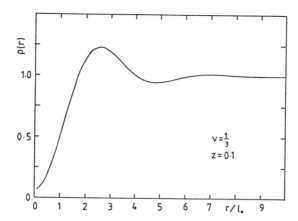

Fig. 6.5. Normalized charge distribution near a repulsive Coulomb impurity with $Z = 0.1$ [6.5]

number on the torus, a topological invariant. Tao and Haldane calculated this chern integral explicitly and found that weak impurities do not change this topological invariant. Lastly, the effect of impurities on the hierarchy of fractional quantum states has been studied by *Zhang* [6.7].

6.2 Higher Landau Levels

In this section, we discuss the FQHE states where higher Landau levels are partially occupied. The earliest study of this problem was by *MacDonald* [6.8]. The higher Landau-level generalization of the Laughlin state is given by

$$|\psi_0^n\rangle = \prod_{i=1}^{N_e} \frac{(a_i^\dagger)^n}{\sqrt{n!}} |\psi_0^0\rangle \qquad (6.12)$$

where $|\psi_0^0\rangle$ is the ground state for $\nu = \frac{1}{m}$, n is the Landau-level index, and a_i is the inter-Landau level ladder operator (see Sect. 5.2). The nth Landau level radial distribution function was then calculated by MacDonald using the plasma analogy and is shown in Fig. 6.6.

In calculating the ground state energy, *MacDonald* and *Girvin* [6.9] noticed that, the relation between pair-correlation function in different Landau levels is much simpler in reciprocal space. The procedure is as follows: a plane wave function of any electron coordinate can be written as

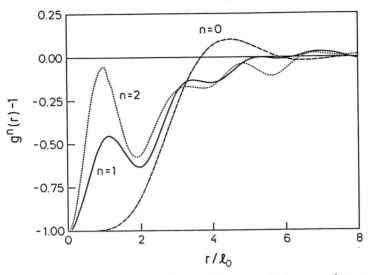

Fig. 6.6. Pair-correlation function for the Laughlin state $g^n(r)$ at $\nu = \frac{1}{3}$ in the case of $n = 0, 1$ and 2 Landau levels [6.8]

$$e^{-i\boldsymbol{k}\cdot\boldsymbol{r}_i} = \exp\left(-\tfrac{1}{\sqrt{2}}ka_i^\dagger\right)\exp\left(-\tfrac{1}{\sqrt{2}}k^*a_i\right)\exp\left(-\tfrac{1}{\sqrt{2}}ikb_i\right)\exp\left(-\tfrac{1}{\sqrt{2}}ik^*b_i^\dagger\right) \tag{6.13}$$

where $a_i, a_i^\dagger$ and $b_i, b_i^\dagger$ are inter-Landau level and intra-Landau level ladder operators respectively, and k is a complex number representation of the wave vector. Recalling the procedure for projection of the density operator onto the nth Landau level (see Sect. 5.2), we have

$$\begin{aligned}\langle\psi_0^n|\rho_{-k}^{nn}\rho_k^{nn}|\psi_0^n\rangle &= \left[L_n\left(\tfrac{1}{2}|k|^2\right)\right]^2\langle\psi_0^n|\bar\rho_{-k}\bar\rho_k|\psi_0^n\rangle \\ &= N_e\left[L_n\left(\tfrac{1}{2}|k|^2\right)\right]^2\bar s(k)\end{aligned} \tag{6.14}$$

where,

$$\bar\rho_k \equiv \sum_i \exp\left[-\tfrac{i}{\sqrt{2}}kb_i\right]\exp\left[-\tfrac{i}{\sqrt{2}}k^*b_i^\dagger\right] \tag{6.15}$$

and $L_n(x)$ is the Laguerre polynomial. For the higher Landau level Laughlin states, it then follows that [6.9]

$$\begin{aligned}h^n(k) &= \rho_0\int d\boldsymbol{r}\, e^{-i\boldsymbol{k}\cdot\boldsymbol{r}_i}[g^n(r)-1] \\ &= \left[L_n\left(\tfrac{1}{2}|k|^2\right)\right]^2 h^0(k).\end{aligned} \tag{6.16}$$

Table 6.1. Energy per particle for Laughlin state and the CDW state in the nth Landau level. The energies are in units of $e^2/\epsilon\ell_0$.

m	$n=0$		$n=1$		$n=2$	
	Laughlin	CDW	Laughlin	CDW	Laughlin	CDW
1	−0.627	−0.627	−0.470	−0.470	−0.401	−0.401
3	−0.409	−0.388	−0.325	−0.316	−0.265	−0.256
5	−0.327	−0.322	−0.294	−0.289	−0.247	−0.250
7	−0.280	−0.279	−0.264	−0.261	−0.252	−0.240
9	−0.250	−0.250	−0.244	−0.238	−0.233	−0.225

In (6.16) ρ_0 is the areal density of electrons. The evaluation of the energy for the state $|\psi_0^n\rangle$ is then

$$N_e^{-1}\langle\psi_0^n|\mathcal{H}|\psi_0^n\rangle = \frac{1}{2}e^2\rho_0 \int d\mathbf{r}\, r^{-1} h^n(r)$$

$$= \frac{1}{2}e^2 \int_0^\infty dq\, [L_n(\tfrac{1}{2}q^2)]^2\, h^0(q). \tag{6.17}$$

Using Laughlin's trial wave function (2.24), the ground state energies for $\nu^{-1} = 3, 5, 7$, and 9 for $n = 0, 1$, and 2 are presented in Table 6.1. The results for the CDW state energies are also given in Table 6.1. They were obtained self-consistently in the HF approximation by MacDonald and Girvin.

From these results, these authors concluded that, (a) the transition to a Wigner crystal ground state occurs at smaller filling factors in higher Landau levels, (b) the incompressible fluid ground state is only marginally stable, if not unstable, in the case where $\nu^{-1} \geq 3 + 2n$ is not satisfied.

These authors have also studied the collective excitations in the higher Landau levels by writing the approximate many-body states as

$$|\psi_k^n\rangle = \frac{\rho_k^{nn}|\psi_0^n\rangle}{[\langle\psi_0^n|\rho_{-k}^{nn}\rho_k^{nn}|\psi_0^n\rangle]^{\frac{1}{2}}}. \tag{6.18}$$

The collective excitation is then obtained from

$$\langle\psi_k^n|\mathcal{H}|\psi_k^n\rangle = \langle\psi_0^n|\mathcal{H}|\psi_0^n\rangle + \Delta^n(k), \tag{6.19}$$

where,

$$\Delta^n(k) = \frac{\langle \psi_0^n | [\rho_{-k}^{nn}, [\mathcal{H}, \rho_k^{nn}]] | \psi_0^n \rangle}{\langle \psi_0^n | \rho_{-k}^{nn} \rho_k^{nn} | \psi_0^n \rangle}. \qquad (6.20)$$

Within the subspace associated with the nth Landau level

$$[\mathcal{H}, \rho_k^{nn}] = \frac{1}{2} \int \frac{d\mathbf{q}}{(2\pi)^2} v(q) \left[L_n \left(\tfrac{1}{2} |q|^2 \right) \right]^2 \left[\bar{\rho}_{-q} \bar{\rho}_q, \bar{\rho}_k \right]. \qquad (6.21)$$

Using (6.21) and (4.47) in (6.20) one obtains,

$$\Delta^n(k) = \int \frac{d\mathbf{q}}{(2\pi)^2} v(q) \left[L_n \left(\tfrac{1}{2} |q|^2 \right) \right]^2 \left[\left(e^{(q^*k - k^*q)/2} - 1 \right) \right.$$
$$\left. \times \bar{s}(q) e^{-\frac{1}{2}|k|^2} + \left(e^{k \cdot q} - e^{k^* q} \right) \bar{s}(k+q) \right] \Big/ \bar{s}(k). \qquad (6.22)$$

The collective excitations obtained from (6.22) for $\nu = \tfrac{1}{3}$ and $\nu = \tfrac{1}{5}$ are plotted in Figs. 6.7 and 6.8 respectively. Let us first consider the results for $\nu = \tfrac{1}{3}$. The curve for $n = 1$, when spin degeneracy and particle-hole symmetry are taken into account, in fact, describes the *total* filling fractions $\nu^* = \tfrac{7}{3}, \tfrac{8}{3}, \tfrac{10}{3}$ and $\tfrac{11}{3}$ in the strong-field limit. The dispersion has lower energy than that for $n = 0$. The result at $\nu = \tfrac{1}{5}$ for $n = 1$ is however quite the opposite, presumably indicating that at $\tfrac{1}{5}$, the system is far from the crystallization transition for the higher Landau level.

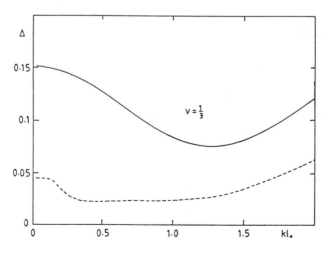

Fig. 6.7. Collective modes at $\nu = \tfrac{1}{3}$ for the $n = 0$ Landau level (solid line) and the $n = 1$ Landau level (dashed line) [6.9]

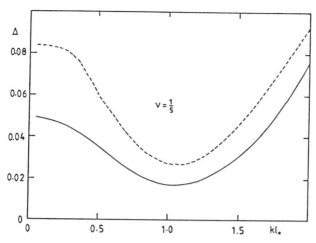

Fig. 6.8. Same as in Fig.6.7, but for $\nu = \frac{1}{5}$ [6.9]

Haldane [6.10] has recently performed a numerical diagonalization of a six-electron Hamiltonian with periodic boundary condition on a hexagonal cell. He obtained a *gapless* ground state at $\nu = \frac{1}{3}$ for $n = 1$. Finite system calculations have also been performed by *d'Ambrumenil* and *Reynolds* [6.11]. Their conclusion is that FQHE is likely to occur at $\nu = \frac{1}{5}$ at $n = 1$. However, at $\nu = \frac{1}{3}$, the Laughlin wave function is not a good candidate for the ground state. A small gap is predicted for this filling.

Experimentally, some tendency to form plateaus at $\nu = \frac{19}{7}$ and $\frac{8}{3}$ has been reported [6.12]. More experimental and theoretical work is needed to clarify the situation.

Finally, estimates for the quasiparticle and quasihole energies for higher Landau levels are also provided by *MacDonald* and *Girvin* [6.13]. The quasiparticle-quasihole energy gap is reported as $E_g(n = 1) = 0.059$ for $\nu = \frac{1}{3}$ and 0.043 for $\nu = \frac{1}{5}$ (in units of $e^2/\epsilon \ell_0$).

6.3 Even Denominator Filling Fractions

In our discussion of the theoretical work so far, we have only described the filling fractions with *odd* denominators. The fact that all investigations focused their attention mostly on these fractions is hardly surprising, because the experimental results clearly demonstrated that for FQHE to occur, the odd denominators are favored exclusively. As we recall, Laughlin's approach *explains* such a fact by the requirement of antisymmetry under interchange of particles and in the hierarchial scheme, such filling factors are taken as the starting point in developing the higher-order filling factors with odd denominators. The possibility of observing FQHE for *even* denominator filling factors is not excluded, however, in these theories.

The simplest filling factor with even denominator is $\nu = \frac{1}{2}$. In this case, the Laughlin-type wave function would describe a system of particles obeying *Bose* statistics.[1] However, one can group the electrons into bound *pairs*, and the pairs can then transform as bosons under interchange of their positions [6.15], and a Laughlin-type wave function could still be used. For the finite-size system in a periodic rectangular geometry (Sect. 2.1), with particles obeying *Bose* statistics, a *cusp* at $\nu = \frac{1}{2}$ was in fact observed by *Yoshioka* [6.16]. In this geometry, collective modes of the type described in Sect. 4.2, were also calculated by *Haldane* [6.17].

The ground state energy (per electron) at $\nu = \frac{1}{2}$, calculated for four- to ten-electron systems in a periodic rectangular geometry is shown in Fig. 6.9. In contrast to the case of $\nu = \frac{1}{3}$, the results in this case show strong dependence on the electron number. The extrapolation of the results in the thermodynamic limit leads to the energy $\approx -0.472\, e^2/\epsilon \ell_0$. The results are of course, lower compared to the crystal energy in the HF limit: The energy difference is ≈ 0.028, while for $\nu = \frac{1}{3}$, the corresponding energy difference is ≈ 0.025. However, the energy difference is much smaller (~ 0.01) for the crystal energies obtained for the four-electron system by Yoshioka et al. (see Fig. 2.2). Given such a small difference, it is not possible to entirely rule out the crystal state at $\nu = \frac{1}{2}$. In particular, the absence of FQHE in the experimental results makes this possibility even more plausible. Improved crystal state calculations are urgently required to settle this interesting issue. The Laughlin state energy at $\nu = \frac{1}{2}$, which corresponds to the boson system, is also given in Fig. 6.9.

The low-lying excitations for several finite-size systems in the periodic rectangular geometry are presented in Fig. 6.10. When we compare these results with the spectrum for an incompressible fluid state (Fig. 4.4), no clear trend is visible in the present spectra. The first difference to note is that, while at $\nu = \frac{1}{3}$, the ground state appears at $k = 0$, in the present case of $\nu = \frac{1}{2}$, the ground state appears at a finite k and varies strongly with the particle number and geometry of the cell (the aspect ratio in these calculations is taken to be $N_e/4$). The whole spectrum is, in fact, particle-number and geometry dependent. The spectrum for an odd number of electrons seems to be somewhat different from that with an even number, presumably indicating the importance of the electron-pair state proposed

---

[1] Interestingly, in a recent article, *Kalmayer* and *Laughlin* have shown that such a wave function would also describe the ground state of a two-dimensional Heisenberg antiferromagnet on a triangular lattice [6.14].

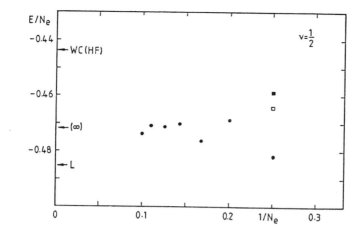

Fig. 6.9. Ground state energy per electron (in units of $e^2/\epsilon\ell_0$) as a function of electron number in a periodic rectangular geometry. The closed and open squares correspond to the crystal energies of Fig.2.2, and the HF energy is given for comparison. The energy of the Laughlin state (depicted as L) and the extrapolation of the finite system results to the thermodynamic limit [depicted as (∞)] are also given

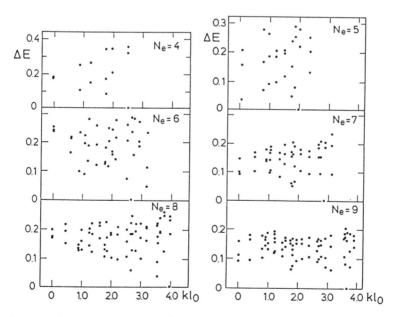

Fig. 6.10. Low-lying excitations at $\nu = \frac{1}{2}$ for finite-size systems in a periodic rectangular geometry as a function of $k\ell_0$

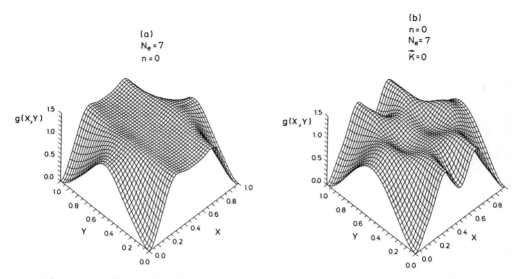

Fig. 6.11. Perspective view of the pair-correlation function $g(r)$ for a seven-electron system at $\nu = \frac{1}{2}$ in the lowest Landau level (a) for the ground state and (b) at $k = 0$, which is an excited state of the system. The axes are normalized as $X = x/a$ and $Y = y/b$

by *Halperin* [6.15]. However, no general conclusion can be drawn from the present numerical results. Although the lowest energy excitations are separated from the higher energy states (more clearly seen for the odd-electron systems) no clear-cut gap structure is apparent in general.

For the half-filled lowest Landau level, *Kuramoto* and *Gerhardts* noticed that, within the HF approximation the ground state is a square CDW with the electron-hole symmetry spontaneously broken [6.18]. The ground state has an energy gap, $\Delta E = 0.330 \, e^2/\epsilon\ell_0$. The density profile was, however, found to be very similar to the self-dual square CDW.

In Fig. 6.11(a), we present the pair-correlation function $g(r)$ [Eq. (2.17)] of the ground state for a seven-electron system at $\frac{1}{2}$ in the lowest Landau level. The function has very little structure and is obviously not isotropic. Qualitatively similar results were obtained for other system sizes [6.19]. In Fig. 6.11(b), we present the correlation function for the same system as in Fig. 6.11(a), but at $k = 0$. The function is clearly isotropic (within the rectangular geometry) and has a liquid-like behavior. From these results, we conclude [6.19] that the half-filled lowest Landau level is *not* a stable state. The translationally invariant liquid state appears as an excited state of the system.

Fano et al. [6.20] have recently done extensive numerical calculations for $\nu = \frac{1}{2}$ in a spherical geometry for up to twelve electrons. They found that, for even number of electrons, the ground state occurs at $L = 0$ only

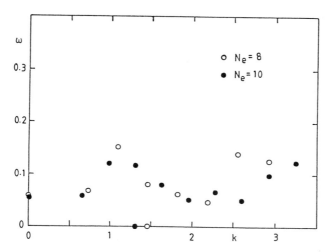

Fig. 6.12. Lowest-lying excitation energies for $N_e = 8$ and 10 as a function of $k = L/R$ in a spherical geometry [6.19]

for $N_e = 6$ and 12. While for other values of N_e, the ground state occurs at, $N_e = 4; L = 2, N_e = 8, 10; L = 4$. For odd numbers of electrons, the ground state has half-integer L. Extrapolating the results for an infinite system, they obtained the ground state energy (per particle) to be, -0.469 ± 0.005.

The excitation spectrum obtained by Fano et al. showed many irregular features, which distinguishes them from the spectrum one gets for an incompressible state at $\nu = \frac{1}{3}$ (see Chap. 4). In Fig. 6.12, we present the lowest energy spectrum for $N_e = 8$ and 10, and in Fig. 6.13, the lowest energy spectrum for $N_e = 6$ and 12. Only in the latter case is the ground state obtained at $L = 0$, however, no clear picture has emerged from these spectra.

Experimentally, the possibility of observing FQHE at even denominator filling factors was indicated by several groups. A minimum in ρ_{xx} at $\nu = \frac{3}{4}$ was first observed by *Ebert* et al. [6.21] in the lowest Landau level. Recently, *Clark* et al. [6.12,22] observed minima in the diagonal resistivity in the second Landau level at $\nu = \frac{9}{4}, \frac{5}{2}$ and $\frac{11}{4}$. Correct quantization of ρ_{xy} to these fractional values were not achieved however.

A thorough analysis of these filling factors has recently been performed by *Willet* et al. [6.23]. Their results for $\nu < 1$ (Fig. 1.3) do not show any sign of FQHE for even denominator fillings. While some features in ρ_{xx} were seen at $\nu = \frac{3}{4}$, two higher order odd denominator filling $\nu = \frac{4}{5}$ and $\nu = \frac{5}{7}$ seem to converge toward this even denominator filling factor. For $\nu = \frac{1}{2}$, ρ_{xy} follows the classical straight line, while the broad minimum in ρ_{xx} is thought to be caused by as yet unresolved higher order odd denominator filling factors.

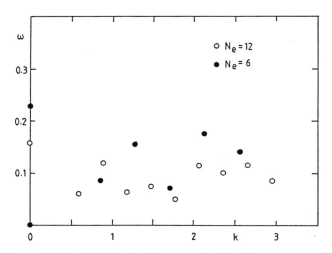

Fig. 6.13. Same as in Fig. 6.12, but for $N_e = 6$ and 12 [6.19]

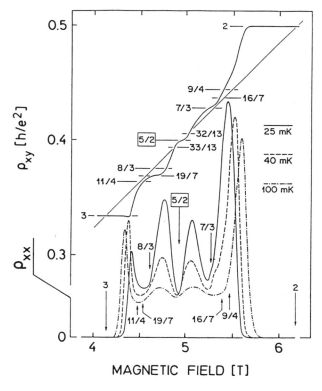

Fig. 6.14. The region of filling factors $3 > \nu > 2$ of Fig. 1.3 shown in detail for temperatures T = 100 to 25 mK [6.23]

For $3 > \nu > 2$ (the Landau level $n = 1$), however, the situation is entirely different. This region of filling factors in Fig. 1.3 is presented in more detail in Fig. 6.14. The ρ_{xy} curve shows a plateau at the magnetic field which corresponds to $\nu = \frac{5}{2}$, which is centered at $\rho_{xy} = (h/e^2)/\frac{5}{2}$ to within 0.5%. In the same region of magnetic field, a deep minimum is observed in ρ_{xx}. The nearest odd-denominator fillings $\frac{32}{13}$ and $\frac{33}{13}$ ($\frac{5}{2} \pm 1.5\%$) do not show any indication that the plateau at $\frac{5}{2}$ is caused by the blending of two higher order odd denominators. There are also broad minima near $\nu = \frac{9}{4}$ and $\frac{11}{4}$ which shift considerably with temperature. However, at the lowest temperature considered in these experiments, these even denominator fractions move to $\frac{19}{7}$ and $\frac{7}{3}$. Therefore, the only even-denominator fraction so far found is $\nu = \frac{5}{2}$. There are indications that other even denominator filling factors exist within the range $4 > \nu > 3$.

There have been a few theoretical attempts to explain the experimental finding discussed above. *Haldane* and *Rezayi* [6.24] recently proposed a spin-singlet state for the incompressible fluid state, which would exhibit a *half-integral* quantum Hall effect. From small-system calculations, they concluded that such a state might be responsible for the $\frac{5}{2}$ effect. Recent numerical calculations by the present authors, also based on the finite-size systems do not support this conclusion however. We predict [6.19] the $\frac{5}{2}$ state to be a fully spin-polarized state.

6.4 Half-Filled Landau Level in Multiple Layer Systems

In the preceding section, we have reviewed the theoretical work on the collective excitations for $\nu = \frac{1}{2}$ in a single layer of electrons. In this section, we present some novel results of the present authors [6.25] for a *layered* electron system. Multilayer electron systems have been studied earlier quite extensively as an anisotropic model for an electron gas [6.26–29]. Let us consider a model where two layers with equal density of electrons are embedded in an infinite dielectric. We consider the delta-function-localized electron density in each plane. The electrons move freely in each plane and the interaction of electrons in different planes is considered to be Coulombic. Tunneling of electrons between the two planes is not allowed. The electrons are also considered to be in their lowest subband. This model is often referred to in the literature as the Visscher-Falicov model [6.29]. Experimental systems that can be described reasonably well by the above model have been obtained in GaAs-heterostructures by different experimental groups [6.30,31].

The Coulomb potential energy between two electrons situated in planes i and j can be written as

$$V(r - r'; i, j) = \frac{e^2}{\epsilon} \left[(r - r')^2 + (i - j)^2 c^2\right]^{-\frac{1}{2}} \tag{6.23}$$

where r is a two-vector (x,y), ϵ is, as usual, the background dielectric constant, and c is the interlayer separation. The Fourier transform of the above expression with respect to $r - r'$ is

$$v(k; i, j) = \frac{2\pi e^2}{\epsilon k} e^{-k|i-j|c}, \tag{6.24}$$

where k is a two-dimensional in-plane wave vector. The effect of interlayer interaction is to lift the two-fold degeneracy which would be present otherwise.

The excitation spectrum for a finite electron system was obtained by generalizing the method of Haldane [see Sect. 4.2] for a two layer system. We consider a rectangular cell consisting of two layers each containing equal number of electrons N_e. For simplicity, we ignore the Landau level mixing and impose periodic boundary conditions such that the cell contains an integer N_s of flux quanta. Furthermore, we consider the electrons to be spin-polarized and take the strong-field limit where the electrons are in the lowest Landau level. The filling fraction is therefore $\frac{1}{2}$ in both the layers. The Hamiltonian now conserves the *total* momentum as well as the number of electrons in each layer. One can, therefore, diagonalize the Hamiltonian for the set of states, $|k_1; L_1\rangle |k - k_1; L_2\rangle$, where $|k_1; L_1\rangle$ is the momentum eigenstate for N_e electrons in a single layer i belonging to the eigenvalue k_i. Here, $L_i = |j_1, \ldots, j_{N_e}\rangle$ labels a Slater determinant of Landau orbitals with momentum k_i.

In Figs. 6.15 and 6.16, we have presented the excitation spectrum for a system with four electrons per layer, for two different values of aspect ratio, $\lambda = 1$ (square) and $\lambda = 1.25$ (rectangular), respectively. The first important result is that the ground state is obtained uniquely at $k = 0$ and it remains so for two different aspect ratios. The ground state energy is, in fact, slightly *lower* for the rectangular geometry. The energy difference between the two geometries is very small however (~ 0.005). The other interesting result is that a gap structure in the spectrum is obtained with a characteristic minimum at a finite $k\ell_0$, similar to that of the magnetoroton minimum, discussed in Sect. 4.4. For the square geometry, a few energy levels lie very close, but are separated from the continuum by a large gap (Fig. 6.15). These close-lying energy states can be further separated by moving away from the square geometry (Fig. 6.16). This is due to the fact that the square geometry, as noted by *Maksym* [6.32], has higher degeneracies as compared to the rectangular geometry.

In Fig. 6.16, the lowest two excitation energies, which are clearly separated from the higher energy states for most values of $k\ell_0$, could presumably

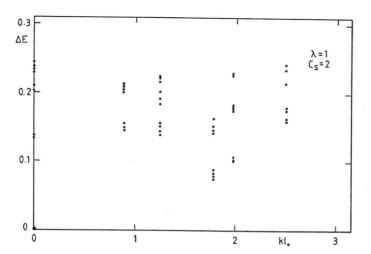

Fig. 6.15. Excitation spectrum of the eight-electron system in a two layer geometry at $\nu = \frac{1}{2}$ for a dimensionless layer separation parameter $C_s = C/\ell_0 = 2.0$ and aspect ratio, $\lambda = 1$ (square cell)

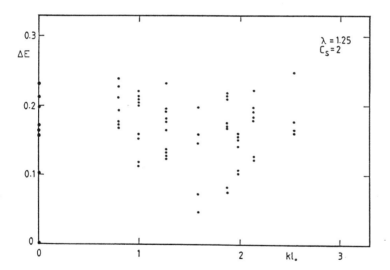

Fig. 6.16. Same as in Fig. 6.15, but for the aspect ratio $\lambda = 1.25$ (rectangular cell)

be interpreted as the two eigenmodes in a system of two charge layers. They arise due to the electron correlations in the two layers. Such a spectrum at $\nu = \frac{1}{3}$ has been obtained in the SMA [6.33] and discussed in Sect. 4.4.

Comparing the layered system results with those of the single layer, discussed in the preceding section, we notice that the introduction of an interacting electron layer has helped to reorganize the excitation energies of the system, particularly the $k=0$ state. The observation of the roton-type minimum is also quite interesting. The results indicate the possibility of the occurrence of an incompressible fluid state at $\nu = \frac{1}{2}$ in a multiple layer system. For a better understanding of the layered system results, we need to obtain spectra for larger systems. As discussed in the previous section, a better understanding of the single-layer spectrum at $\nu = \frac{1}{2}$ is also very much needed. Experimental investigations of FQHE in a layered system, such as the one considered above, would be very interesting.

7. Open Problems and New Directions

Before we close the review, a brief recapitulation of some of the open problems in this field is perhaps in order. Despite the fact that the Laughlin-type approaches, as discussed in this review, are quite successful in describing many important aspects of FQHE, the physical origin of the correlations that are described by the present approaches is not completely clear. Independent approaches, like the cooperative-ring-exchange mechanism might be helpful in providing an answer to that problem. From our point of view, the latter approach should be allowed to generate some quantitative results which are comparable to the theoretical results already established by various approaches and described in this review.

On a more practical level, the important problem of finding the critical filling factor at which the liquid to solid transition occurs is to be studied further. This would require improved calculations for the crystal state, as well as further experimental investigations at low densities.

In the case of elementary excitations, our understanding is still far from complete. Much theoretical work is needed to get the quasiparticle energy in the Laughlin-type approach to match with the energy gap obtained via the finite-size calculations, as well as the Monte Carlo estimates. This would be a very important step for calculation of the collective excitation spectrum in Laughlin's theory. Further theoretical work is also needed to remove the discrepancy between the theoretical and experimental results for the energy gap. Influence of disorder on the energy gap is a very important problem to be addressed theoretically. Attempts should also be made to observe the collective modes experimentally. While the theoretical work on the collective excitation spectrum at $\nu = \frac{1}{3}$ is quite exhaustive, there have only been a few attempts to study the higher order filling fractions. Similarly, the inter-Landau level collective mode in the case of fractional filling of the lower Landau level, and the possible role of impurities on this mode need to be studied further.

While Laughlin's wave function is very successful in providing a quantitatively correct picture for $\frac{1}{m}$ filling factors (and $1 - \frac{1}{m}$ with particle-hole symmetry), we have not reached such a stage with the higher order fillings. Further theoretical investigations of the experimentally observed higher order filling factors are very much needed. Attempts have to be made to obtain quantitatively accurate results for these states, both in the hierarchial scheme as well as in the trial wave function approach.

Perhaps the most interesting *test* of Laughlin's wave function will be in the case of *even*-denominator filling factors. Whether the experimentally observed $\frac{5}{2}$ fraction describes a different state or is still a Laughlin-type state remains to be investigated further. The ground state and the excitations for the even-denominators, as we have tried to demonstrate in Sect. 6.3, are still unclear. The numerical diagonalization scheme described in Sect. 6.3, has shown that the state is not an incompressible fluid. The experimental finding of one of the even denominators has made the problem quite pressing for the theorists. Layered system results, as discussed in Sect. 6.4, perhaps provide a new opportunity to observe the even denominator fractions. More theoretical and experimental work is needed to settle these important issues.

Appendices

A The Landau Wave Function in the Symmetric Gauge

In this appendix, we derive the eigenstates of a single electron in the presence of a uniform magnetic field, with the choice of *symmetric gauge* vector potential, $A = \frac{1}{2}B(y\hat{x} - x\hat{y})$. This choice of vector potential was adopted by Laughlin for his theory described in Sect. 2.2. The single-electron eigenstates are given explicitly by *Dingle* [A.1] (also see [A.2]), in cylindrical polar coordinates. In the following, we present an alternative derivation of the same problem.

In terms of the dimensionless complex coordinates,

$$z = \frac{1}{2\ell_0}(x - iy)$$
$$z^* = \frac{1}{2\ell_0}(x + iy), \qquad \text{(A.1)}$$

the Hamiltonian (2.1) is written as

$$\mathcal{H} = \frac{1}{2}\hbar\omega_c \left[\left(-i\ell_0 \partial_x - \frac{y}{2\ell_0}\right)^2 + \left(-i\ell_0 \partial_y + \frac{x}{2\ell_0}\right)^2\right]$$
$$= \tfrac{1}{2}\hbar\omega_c \left[z^* z + z^* \partial_{z^*} - z\partial_z - \partial_z \partial_{z^*}\right] \qquad \text{(A.2)}$$

Let us now define the operators,

$$a^\dagger = \frac{1}{\sqrt{2}}(z^* - \partial_z)$$
$$a = \frac{1}{\sqrt{2}}(z + \partial_{z^*}),$$

with the commutation relation $[a, a^\dagger] = 1$. The Hamiltonian is then simply

$$\mathcal{H} = \tfrac{1}{2}\hbar\omega_c \left(a^\dagger a + a a^\dagger\right)$$
$$= \hbar\omega_c \left(a^\dagger a + \tfrac{1}{2}\right) \qquad \text{(A.3)}$$

as expected. The Landau levels are then obtained by diagonalizing the Hamiltonian.

Let us now introduce another set of operators b and $b^\dagger$, which commute with a and $a^\dagger$ and have the same commutation relations as the a's:

$$\begin{aligned} b &= \frac{1}{\sqrt{2}}(z + \partial_{z^*}) \\ b^\dagger &= \frac{1}{\sqrt{2}}(z^* - \partial_z). \end{aligned} \qquad (A.4)$$

The eigenstates of the Hamiltonian are now chosen such that

$$\begin{aligned} \mathcal{H}\psi_{mn} &= \hbar\omega_c\left(n + \tfrac{1}{2}\right)\psi_{mn} \\ b^\dagger b\,\psi_{mn} &= m\,\psi_{mn}, \end{aligned} \qquad (A.5)$$

where m and n are integers with n being the Landau level index and m is the degeneracy of that Landau level. Let us now define the lowest states by

$$a\psi_0 = (z^* + \partial_z)\psi_0 = 0, \qquad (A.6)$$

which then leads to

$$\psi_0 = C e^{-z^* z} \qquad (A.7)$$

with C as a constant.

The higher states are defined accordingly by

$$\begin{aligned} \psi_{mn} &= C\left(b^\dagger\right)^m \left(a^\dagger\right)^n \psi_0 \\ &= C\left(z^* - \partial_z\right)^m (z + \partial_{z^*})^n e^{-z^* z}. \end{aligned} \qquad (A.8)$$

For $n = 0$, we obtain the states corresponding to the lowest Landau level,

$$\begin{aligned} \psi_{0m} &= C\left(z^* - \partial_z\right)^m e^{-z^* z} \\ &= C z^m e^{-z^* z}. \end{aligned} \qquad (A.9)$$

In (A.9), we have made use of the operator identity,

$$z^* - \partial_z = -e^{z^* z}\left(\partial_z\right)e^{-z^* z}. \qquad (A.10)$$

The constant in (A.8,9) is obtained from the normalization condition. Now choosing the conventional form of complex coordinates which we have used in Sect. 2.2, $z = x - iy$, the single-particle wave function in the lowest Landau level is obtained as

$$\psi_{0m} = \frac{1}{(2\pi\ell_0^2 2^m m!)^{\frac{1}{2}}}\left(\frac{z}{\ell_0}\right)^m e^{-|z|^2/4\ell_0^2}. \qquad (A.11)$$

It is easy to verify that, ψ_{0m} is also an eigenfunction of the angular momentum operator $L_z = -i\hbar(x\partial_y - y\partial_x)$ with eigenvalue m.

It is often convenient to define the ladder operators following *Girvin* and *Jach* [A.3,4]. The lowest Landau level eigenfunction obtained above is now written as (in units where $\ell_0^{-2} = eB/\hbar c = 1$):

$$\psi[z] = f[z]\, e^{-\frac{1}{4}\sum_i |z_i|^2} \tag{A.12}$$

where $[z] \equiv (z_1, \ldots, z_N)$ and f is a polynomial in the N variables. The exponential factor in (A.12) is common to all wave functions, and is removed by defining a suitable Hilbert space of analytic functions [A.5] as described below.

Following [A.3,4], let us now consider a set of entire functions of N complex variables, $\Theta \equiv \{f\}$. These functions are analytic in each of their arguments in the complex plane. For example, for $N = 1$, the function $f(z) = z^3$ is an element of Θ, but the function $f(z) = z^*$ is *not* analytic, (as z^* cannot be expressed as a power series in z), and is excluded from Θ.

Let us now define an inner product on Θ as

$$(f, g) = \int d\mu[z]\, f^*[z]\, g[z], \tag{A.13}$$

with the measure

$$d\mu[z] = \prod_{i=1}^{N} \frac{1}{2\pi} e^{-|z_i|^2/2}\, dx_i\, dy_i. \tag{A.14}$$

Only those functions with finite norm $(f, f) < \infty$ are included in Θ. The Hilbert space as defined above, is realized by the wave functions of the lowest Landau level as they may always be written in the form of (A.12) with f being a member of Θ. The inner product on Θ is defined in such a manner that

$$\langle \psi' | \psi \rangle = (f', f). \tag{A.15}$$

This definition of the Hilbert space allows one in (A.12) to work only with f, which is analytic, while ψ is not.

Let us now define the orthonormal basis functions,

$$f_n(z) = \frac{z^n}{(2^n n!)^{\frac{1}{2}}}, \tag{A.16}$$

where the following relations can easily be verified:

$$z f_m = \sqrt{2}\sqrt{m+1}\, f_{m+1}$$
$$\frac{d}{dz} f_m = \sqrt{\frac{m}{2}}\, f_{m-1}.$$
(A.17)

Therefore one can define the boson ladder operators [A.3,4] as:

$$a^\dagger \equiv \frac{1}{\sqrt{2}} z$$
$$a \equiv \sqrt{2}\, \frac{d}{dz}$$
(A.18)

which are mutually adjoint with respect to the inner product defined in Θ.

It is interesting to observe at this point that the adjoint of z is not the same as the Hermitian conjugate of z. From the definition of the inner product, it is clear that

$$(f_n, z_k^* f_m) = (z_k f_n, f_m)$$
(A.19)

as z_k^* is the Hermitian conjugate of z_k. However, from (A.18), the adjoint of z_k is

$$z_k^\dagger = 2\frac{\partial}{\partial z_k}$$
(A.20)

so that

$$(f_n, z_k^* f_m) = \left(f_n, 2\frac{\partial}{\partial z_k} f_m\right).$$
(A.21)

Therefore, z_k^* and $2\partial/\partial z_k$ have the same matrix elements within the space Θ. However the two operators are not completely equivalent since z_k^* commutes with z_k but $\partial/\partial z_k$ does not. As an example,

$$(f, z_k z_k^* g) = (f, z_k^* z_k g)$$
(A.22)

but

$$\left(f, z_k 2\frac{\partial}{\partial z_k} g\right) \neq \left(f, 2\frac{\partial}{\partial z_k} z_k g\right).$$
(A.23)

It is clear that only the right-hand side of (A.23) agrees with (A.22). These observations were useful in obtaining the projection of various operators discussed in Sect. 4.4.

B The Hypernetted–Chain Primer

The hypernetted-chain (HNC) method is a well-established procedure to calculate the pair-correlation function for an imperfect gas in classical statistical mechanics. The method was originally proposed by *van Leeuwen* et al. [B.1] and independently by *Morita* and *Hiroike* [B.2]. In this appendix, we present a brief discussion of the method following closely the paper by van Leeuwen et al.

The pair-correlation function at temperature $T = (\beta k_B)^{-1}$, and density ρ is given by

$$g(r_{12}) = \frac{N(N-1)}{\rho^2} \frac{\int \exp\left[-\beta \sum_{i<j} \varphi(r_{ij})\right] dr_3 \ldots dr_N}{\int \exp\left[-\beta \sum_{i<j} \varphi(r_{ij})\right] dr_1 \ldots dr_N} \tag{B.1}$$

where $\varphi(r_{ij})$ is the interaction between particles i and j. Let us define the functions

$$f(r_{ij}) = \exp\left[-\beta \varphi(r_{ij})\right] - 1 \tag{B.2}$$

$$h(r_{12}) = g(r_{12}) - 1. \tag{B.3}$$

For values of r_{12} much larger than the range of interparticle interaction, the pair-correlation function is rigorously equal to 1 and hence, $h(r_{12})$ is a short range function, $h(r_{12}) \to 0$ for large r.

One can write the correlation function as a series expansion in powers of density as

$$h(r_{12}) = \exp\left[-\beta \varphi(r_{12})\right] \cdot \left[1 + C(r_{12})\right] - 1, \tag{B.4}$$

where,

$$\begin{aligned} C(r_{12}) &= \sum_{k=1}^{\infty} \rho^k \gamma_k(r_{12}) \\ &= \sum_{k=1}^{\infty} \frac{\rho^k}{k!} \int \ldots \int \sum_{(\text{sp.irr.})} \prod f(r_{ij}) \, dr_3 \ldots dr_{k+2}. \end{aligned} \tag{B.5}$$

The summation inside the integral of (B.5) is over all different products of $f(r_{ij})$ [excluding $f(r_{12})$] corresponding to the *specific* (distinguishable) *irreducible* $1-2$ diagrams to be described below.

Before we proceed with the diagrammatic scheme, let us first explain the terminologies involved. We define the *external* (or *reference*) points as coordinates which are not integration variables (points 1 and 2) and the *internal* (or *field*) points as the coordinates which are integration variables

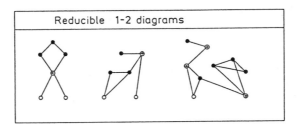

Fig. B.1. Some examples of reducible $1-2$ diagrams. The articulation points are drawn as $\oplus$

(points $3,\ldots,k+2$). The internal points are connected by a certain number of bonds, each bond corresponding to one of the factors $f(r_{ij})$ in the product of (B.5). An *irreducible* diagram is *linked* or *connected* (not containing a direct bond $1-2$) in the sense that each point is connected to every other point by at least one bond, and has no *articulation points*. An articulation point is a point where the connected diagram may be split into separate connected subdiagrams.[1] The external points can also be articulation points (Fig. B.1).

In (B.5), the contributions to the integral from two specific diagrams which differ only through a permutation of k internal points, are the same. Therefore, one could perform the summation in (B.5) over the different *generic* diagrams or different topological types of diagrams. The total number of specific diagrams which correspond to one particular generic diagram is $k!/s$ where s is the symmetry number denoting the number of permutations that do *not* lead to a new specific diagram. Then (B.5) is rewritten as

$$C(r_{12}) = \sum_{k=1}^{\infty} \frac{\rho^k}{k!} \int \cdots \int \sum_{\substack{\kappa \\ \text{(gen.irr.)}}} \frac{1}{s(k,\kappa)} \prod f(r_{ij})\, dr_3 \ldots dr_{k+2} \quad (\text{B.6})$$

where the summation runs over all different generic irreducible $1-2$ diagrams κ with k internal points and $s(k,\kappa)$ is the symmetry number corresponding to the generic diagram κ. In (B.6), $C(r_{12})$ is thus a sum of contributions due to all possible generic irreducible $1-2$ diagrams in which the contribution of a particular generic diagram is equal to the product of ρ^k; $1/s(k,\kappa)$ accounts for the symmetry, and the integral corresponding to that diagram.

---

[1] An $i-j$ *subdiagram* is a part of $1-2$ diagram which is only connected with the rest of the diagram through points i and j.

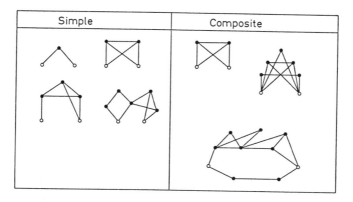

Fig. B.2. Some examples of simple and composite irreducible $1-2$ diagrams (the C-set)

Parallel connection of diagrams: An irreducible $1-2$ diagram might be composed of two or more $1-2$ *sub*diagrams forming parallel connections between the points 1 and 2. As the points 1 and 2 are not integrated over, the integral corresponding to such a graph is factorizable into products of each $1-2$ subdiagram. A diagram of this type is called a *composite* diagram. If not composite, the irreducible $1-2$ diagram is called *simple*. Examples of simple and composite diagrams are given in Fig. B.2.

The contribution from all simple irreducible $1-2$ diagrams with l internal points is written as

$$\beta_l(r_{12}) = \frac{1}{l!} \int \cdots \int dr_3 \ldots dr_{l+2} \sum_{\text{(sp.irr.)}} \prod f(r_{ij})$$

or,

$$\beta_l(r_{12}) = \sum_{\substack{\lambda \\ \text{(gen.simp.)}}} \frac{1}{s(l,\lambda)} \int \cdots \int dr_3 \ldots dr_{l+2} \prod f(r_{ij}).$$

The complete set of all generic composite and simple diagrams of k internal points composed of $\{m_l\} = m_1, m_2, \ldots$ simple subdiagrams of $1, 2, \ldots$ internal points respectively can be obtained as follows: We choose all possible generic simple subdiagrams, one out of each of the $m_1, m_2, \ldots$ complete subset of simple diagrams with $1, 2, \ldots$ internal points. *van Leeuwen* et al. [B.1] then asserted that, in terms of this complete set of simple and composite diagrams, one can write,

$$1 + C(r_{12}) = \sum_{\{m_l\}} \prod_l \frac{1}{m_l!} \left[\beta_l \rho^l\right]^{m_l} \tag{B.7}$$

taking proper account of the symmetry factor and correct counting of the subdiagrams. For the C-set of all composite and simple $1-2$ diagrams the following relation is then obtained,

$$C(r_{12}) = e^{S(r_{12})} - 1 \tag{B.8}$$

with

$$S(r_{12}) = \mathcal{S}f(r) = \sum_{k=1}^{\infty} \beta_k(r_{12})\rho^k \tag{B.9}$$

being the contribution from the $\mathcal{S}$-set of all simple $1-2$ diagrams only. Expanding the exponential we obtain $S(r_{12})$ plus the sum of all composite diagrams with the correct weights. A similar expression was also obtained by *Salpeter* [B.3]. In terms of $C(r_{12})$, the pair correlation function is given by

$$h(r_{12}) = \exp\left[-\beta\varphi(r_{12}) + S(r_{12})\right] - 1. \tag{B.10}$$

Distinguishing in this way between simple and composite diagrams, $C(r_{12})$ can be expressed in terms of the set of simple $1-2$ diagrams.

Series connection of diagrams: The $1-2$ diagram may be either *nodal* or *non-nodal*. A diagram is nodal if it has one or more nodes. A *node* is a point through which all paths from 1 to 2 must go. A composite diagram is by definition non-nodal. Conversely, a nodal (irreducible) $1-2$ diagram is necessarily simple. A simple diagram contributing to $C(r_{12})$ can be either nodal or non-nodal; if non-nodal such a diagram is called *elementary*. Therefore, the contribution $S(r_{12})$ due to all simple $1-2$ diagrams can be divided into contributions $N(r_{12}) = \mathcal{N}f(r)$, due to the $\mathcal{N}$-set of nodal diagrams and $E(r_{12}) = \mathcal{E}f(r)$ due to the $\mathcal{E}$-set of elementary diagrams,

$$S(r_{12}) = N(r_{12}) + E(r_{12}), \tag{B.11}$$

and the pair correlation function is rewritten as

$$h(r_{12}) = \exp\left[-\beta\varphi(r_{12}) + E(r_{12}) + N(r_{12})\right] - 1. \tag{B.12}$$

Examples of elementary diagrams are given in Fig. B.3.

In general, any nodal diagram can be built by connecting in series, a number of non-nodal subdiagrams. Our task is to express the contributions due to the nodal diagrams $N(r_{12})$, by series connection of elements of a set of non-nodal diagrams. We are immediately faced with the problem that the set of non-nodal diagram contained in the C-set — considered as $i-j$ subdiagrams rather than $1-2$ diagrams — are simply not sufficient. For

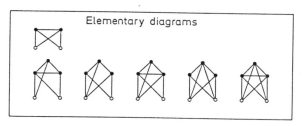

Fig. B.3. Some examples of elementary $1-2$ diagrams (the $\mathcal{E}$-set)

example, an essential element for the construction of nodal diagrams by *series* connection is the direct bond, which is not included in the $\mathcal{C}$-set.

van Leeuwen et al. then introduced an *extended* set of $1-2$ diagrams, the $\mathcal{G}$-set, which included the original $\mathcal{C}$-set, plus all the diagrams obtained by adding the direct bond corresponding to $f(r_{12})$ to the $\mathcal{C}$-set and the direct bond itself. This set contains all the diagrams corresponding to the different terms in $h(r_{12})$:

$$h(r_{12}) = [1 + f(r_{12})] \, C(r_{12}) + f(r_{12}) = \mathcal{G}f(r). \tag{B.13}$$

The $\mathcal{N}$-set of nodal diagrams, being a subset of the original $\mathcal{C}$-set of (composite and simple) diagrams, is also a subset of the (extended) $\mathcal{G}$-set of diagrams. If $X(r_{12})$ denotes the contributions to the $\mathcal{X}$-set of diagrams (complete set of all possible diagrams to be used as a subdiagram in constructing the nodal diagrams) then,

$$X(r_{12}) = h(r_{12}) - N(r_{12}) = \mathcal{X}f(r). \tag{B.14}$$

Using (B.12), one can then write

$$X(r_{12}) = \exp\left[-\beta\varphi(r_{12}) + N(r_{12}) + E(r_{12})\right] - N(r_{12}) - 1. \tag{B.15}$$

Examples of nodal and non-nodal diagrams of the extended set (the $\mathcal{G}$ set) are given in Fig. B.4.

Following van Leeuwen et al., we now derive an integral equation for the quantity $N(r_{12})$ (corresponding to the nodal diagrams of the $\mathcal{G}$-set) in terms of $X(r_{12})$ (which corresponds to the non-nodal diagrams of the $\mathcal{G}$-set):

$$N(r_{12}) = \rho \int \left[X(r_{13}) + N(r_{13})\right] X(r_{32}) dr_3 \tag{B.16}$$

which is completely general and does not involve any approximation. It is

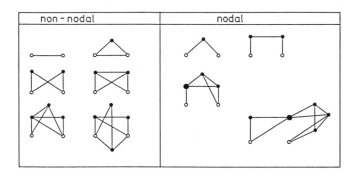

Fig. B.4. Some examples of the nodal and non-nodal diagrams of the extended set (the $\mathcal{G}$-set)

easy to check this equation by iteration. As the integration in (B.16) is of convolution type, going over to the Fourier space, we obtain an algebraic equation,

$$\widetilde{N}(k) = \frac{\rho \left[\widetilde{X}(k)\right]^2}{1 - \rho \widetilde{X}(k)}. \tag{B.17}$$

Again, as in (B.8), a massive partial summation has been achieved.

The pair-correlation function is now finally determined within the hyper-netted-chain scheme by solving the following set of coupled equations which are exact,[2]

$$X(r_{12}) = h(r_{12}) - N(r_{12})$$
$$\widetilde{N}(k) = \rho \left[\widetilde{X}(k)\right]^2 / \left[1 - \rho \widetilde{X}(k)\right] \tag{B.18}$$
$$g(r_{12}) = \exp\left[-\beta\varphi(r_{12}) + E(r_{12}) + N(r_{12})\right].$$

In the actual calculations, however, one faces the problem that the contribution corresponding to all elementary (simple non-nodal) diagrams is not explicitly available as a function of r. As a first approximation, we set $E(r) = 0$. This approximation has been used throughout in the text. To get some feeling about the above equations, we now perform a few iteration steps analytically: In the zeroth order let us put

$$N_0^{(0)} = 0.$$

---

[2] In (2.33), we have defined $C(r_{12}) \equiv X(r_{12})$, used the approximation $E(r_{12}) = 0$ and the density ρ has been absorbed in the Fourier transform, (2.32). Also, the interaction potential is $\beta\varphi(r) = u(r)$.

Then from (B.15) we obtain,

$$X_0^{(0)}(r) = \exp\left[-\beta\varphi(r)\right] - 1 = f(r)$$

(the subscript 0 is to denote the case of $E(r) = 0$). The pair correlation function is then simply

$$h_0^{(0)}(r_{12}) = f(r_{12}) = \exp\left[\beta\varphi(r_{12})\right] - 1.$$

In the first order, we take the Fourier transform, $\widetilde{X}_0^{(0)}(k) = \widetilde{f}(k)$ and obtain

$$\widetilde{N}_0^{(1)}(k) = \frac{\rho\left[\widetilde{X}_0^{(0)}(k)\right]^2}{1 - \rho\widetilde{X}_0^{(0)}(k)} = \frac{\rho\left[\widetilde{f}(k)\right]^2}{1 - \rho\widetilde{f}(k)}$$

and the inverse Fourier transform provides $N_0^{(1)}(r)$, which corresponds to *all possible* nodal diagrams obtained by connecting in series the single bonds $f(r)$. The improved approximation to $X(r_{12})$ is then,

$$X_0^{(1)}(r) = [1 + f(r)]\exp\left[N_0^{(1)}(r)\right] - N_0^{(1)}(r) - 1$$

which corresponds to all possible diagrams obtained by connecting in parallel all possible nodal diagrams out of the set of diagrams which contribute to $N_0^{(1)}(r)$. The pair correlation function, in this iteration step is

$$h_0^{(1)}(r_{12}) = \exp\left[\beta\varphi(r_{12}) + N_0^{(1)}(r_{12})\right] - 1.$$

In the next step, we take the Fourier transform of $X_0^{(1)}(r)$ and evaluate a better approximation,

$$\widetilde{N}_0^{(2)}(k) = \frac{\rho\left[\widetilde{X}_0^{(1)}\right]^2}{1 - \rho\widetilde{X}_0^{(1)}(k)}.$$

The inverse Fourier transform, $N_0^{(2)}(r)$ now corresponds to all possible nodal diagrams connecting in series all possible diagrams out of the set corresponding to $X_0^{(1)}(r)$.

Repeating this procedure of series and parallel connections, starting with the single bond, we generate a huge class of diagrams. In Fig. B.5, we have sketched the iteration procedure, and in Fig. B.6, we show the type of

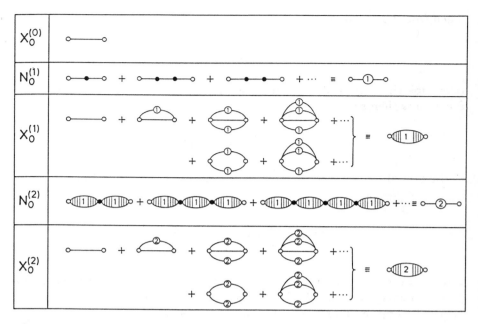

Fig. B.5. The steps of the series-parallel iteration process demonstrated symbolically

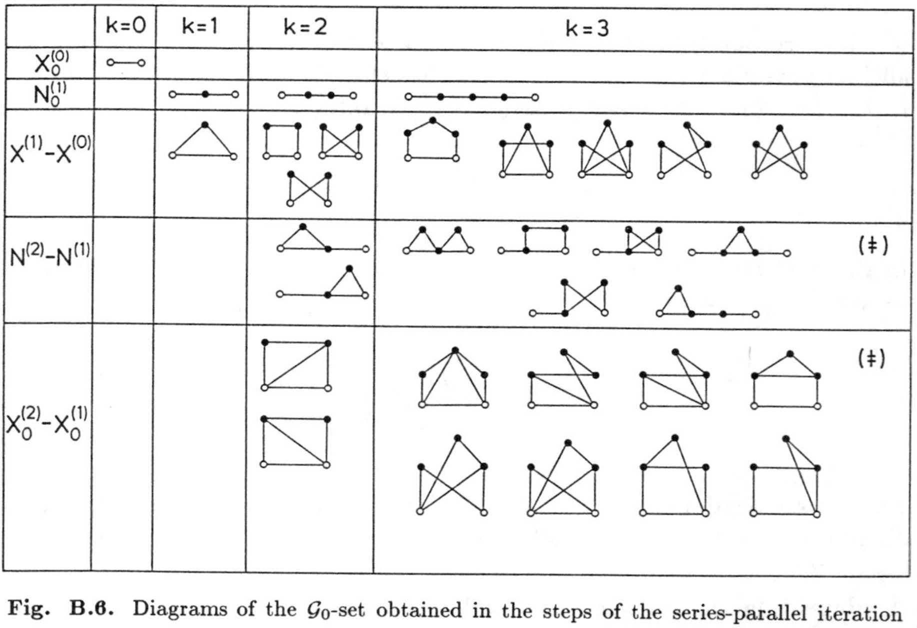

Fig. B.6. Diagrams of the $\mathcal{G}_0$-set obtained in the steps of the series-parallel iteration procedure with 1, 2 and 3 internal points. (‡) The diagrams which one can obtain by an interchange of the external points are omitted

diagrams that are obtained in the successive steps of the iteration procedure. It should be emphasized here that, in each step of the calculation, an infinite set of diagrams up to *infinite order* (but of certain type) are taken into account.

In the case of a classical plasma, the contribution from the elementary diagrams has been approximately evaluated by *Caillol* et al. [B.4]. Making use of the exact results for the correlation function in the case of $\Gamma = 2$, the functions corresponding to elementary diagrams are obtained as:

$$E_{\Gamma=2}(x) = e^{-x^2} + \ln\left[\left(1 - e^{-x^2}\right)/x^2\right] + C(x)$$
$$C_{\Gamma=2}(x) = -2\gamma - \sum_{j=1}^{\infty} \frac{1}{j}\left[e^{-x^2/j} - 1\right] \quad (B.19)$$

where γ is Euler's constant. The function $E(x)$ is then represented quite accurately by the simple functional form,

$$E_{\Gamma=2}(x) = e^{-x^2}\left[E_0 + E_2 x^2 + E_4 x^4 + E_6 x^6\right]$$

with the coefficients: $E_0 = 1 - 2\gamma = -0.1544313$, $E_2 = -0.00949726$, $E_4 = 0.00835662$ and $E_6 = 0.00108723$. Caillol et al. then assumed that the shape of the function $E(x)$ does not change significantly with Γ, and made the following ansatz:

$$E_\Gamma(x) = \alpha \frac{1}{2}\Gamma E_{\Gamma=2}(x). \quad (B.20)$$

The coefficient α is adjusted to achieve thermodynamic consistency.

In the FQHE calculations, our experience is that the pair-correlation functions within the HNC scheme (not including the elementary diagrams) are almost indistinguishable from the very accurate Monte Carlo results. The contribution of the elementary diagrams is noticeable only in the results for energy.

Finally, the two-component HNC method is a straightforward generalization of the single-component HNC procedure discussed above [B.5–7]:

$$g_{\alpha\beta}(x) = \exp\left[N_{\alpha\beta}(x) - u_{\alpha\beta}(x)\right]$$
$$\tilde{N}_{\alpha\beta}(q) = \sum_{\nu=1,2} C_{\nu\alpha}\left[C_{\nu\beta} + N_{\nu\beta}\right] \quad (B.21)$$
$$C_{\alpha\beta}(x) = g_{\alpha\beta}(x) - 1 - N_{\alpha\beta}(x)$$

where $\alpha, \beta = 1, 2$ denotes the two species of particles in the system. The equation for $\tilde{N}_{\alpha\beta}(q)$ is easily solvable and one obtains,

$$\tilde{N}_{11} = \left(\tilde{C}_{11}^2 + \tilde{C}_{12}^2/D\right)/(1-\tilde{C}_{11})$$
$$\tilde{N}_{22} = \left(\tilde{C}_{22}^2 + \tilde{C}_{12}^2/D\right)/(1-\tilde{C}_{22}) \qquad \text{(B.22)}$$
$$\tilde{N}_{12} = \tilde{C}_{12}(1-D)/D$$

with, $D = (1-\tilde{C}_{11})(1-\tilde{C}_{22}) - \tilde{C}_{12}^2$. An iteration scheme similar to that used in the single-component case could be used here to obtain $g_{\alpha\beta}(x)$.

C Repetition of the Intra-Mode in the Inter-Mode

In the following, we shall consider only those kind of inter-Landau level excitations where exactly one electron from the lowest level is elevated to the second level. Our aim is to show that in the spectrum of those excitations there are points which exactly repeat —when shifted by $\hbar\omega_c$— the intra-Landau level spectrum of the lowest Landau level. We shall present the proof for the finite-size system and also provide the evidence for the infinite system within the framework of the SMA model described in Sect. 4.4.

We will first concentrate on systems of finite number of particles. For the time being it is enough if we consider only the potential energy part $\mathcal{V}$ of the total Hamiltonian $\mathcal{H} = \mathcal{K} + \mathcal{V}$ since the kinetic energy $\mathcal{K}$ is diagonal in our representation. In occcupation number space the potential energy part of the Hamiltonian is given by [sec (4.28)],

$$\mathcal{V} = \sum_{\{j_i K_i\}} A_{j_1 K_1 j_2 K_2 j_3 K_3 j_4 K_4} a^\dagger_{j_1 K_1} a^\dagger_{j_2 K_2} a_{j_3 K_3} a_{j_4 K_4}. \qquad \text{(C.1)}$$

To simplify the appearance of the formulas we use the shorthand notation

$$V_{j_1 K_1 j_2 K_2 j_3 K_3 j_4 K_4} = A_{j_1 K_1 j_2 K_2 j_3 K_3 j_4 K_4} a^\dagger_{j_1 K_1} a^\dagger_{j_2 K_2} a_{j_3 K_3} a_{j_4 K_4} \qquad \text{(C.2)}$$

for the terms of the potential energy. The vector

$$|L\rangle = |j_1 0, j_2 0, \ldots, j_{N_e} 0\rangle \qquad \text{(C.3)}$$

is used to describe the state where all N_e electrons are in the lowest Landau level and

$$|L; k\rangle = |j_1 0, j_2 0, \ldots, j_k 1, \ldots, j_{N_e} 0\rangle \qquad \text{(C.4)}$$

the state where exactly one electron is elevated to the next level. The set of corresponding indices $\{j_1, j_2, \ldots, j_{N_e}\}$ is denoted by L. The projections of the Hamiltonian (C.1) onto the subspaces spanned by the vectors (C.3) and (C.4) are denoted by $\mathcal{V}^0$ and $\mathcal{V}^1$, respectively. The shorthand notation (C.2) is applied to the representations of these operators with the obvious usage of superscripts 0 and 1.

To show the existence of the repeated spectrum we choose an arbitrary eigenstate $|\Psi\rangle$ of $\mathcal{V}^0$

$$\mathcal{V}^0|\Psi\rangle = \varepsilon|\Psi\rangle \tag{C.5}$$

where the state $|\Psi\rangle$ is a superposition

$$|\Psi\rangle = \sum_L c_L |L\rangle \tag{C.6}$$

of state vectors of the lowest level. The operator $\mathcal{V}^0$ being a projection onto states (C.3) maps those vectors into a superposition of the same vectors

$$\mathcal{V}^0|L\rangle = \sum_{L'} h_{L'L} |L'\rangle \tag{C.7}$$

where the coefficients are given by

$$h_{L'L} = A_{j_1 0 j_2 0 j_3 0 j_4 0}, \quad \{j_1, j_2\} \subset L', \quad \{j_3, j_4\} \subset L. \tag{C.8}$$

With this notation the eigenvalue equation (C.5) can be written in the form

$$\sum_{L'} h_{L'L} c_{L'} = \varepsilon c_L. \tag{C.9}$$

As a first step of our proof let us show that the state

$$|L\rangle_e = \sum_{k=1}^{N_e} |L; k\rangle \tag{C.10}$$

behaves under the operation of the partial Hamiltonian V^1 according to the formula

$$\sum_{\substack{K_1 K_2 \\ K_3 K_4}} V^1_{i_1 K_1 i_2 K_2 i_3 K_3 i_4 K_4} |L\rangle_e = h_{L'L} |L'\rangle_e . \tag{C.11}$$

Because, by definition, V^1 preserves exactly one electron on the second Landau level, it is easy to see that

$$\sum_{\substack{K_1 K_2 \\ K_3 K_4}} V^1_{i_1 K_1 i_2 K_2 i_3 K_3 i_4 K_4} |L\rangle_e = \sum_{\substack{k \\ j_k \neq i_3 \\ j_k \neq i_4}} V_{i_1 0 i_2 0 i_3 0 i_4 0} |j_1 0, j_2 0, \ldots, j_k 1, \ldots, j_{N_e} 0\rangle$$

$$+ V_{i_1 1 i_2 0 i_3 0 i_4 1} |j_1 0, j_2 0, \ldots, i_4 1, \ldots, j_{N_e} 0\rangle$$

$$+ V_{i_1 0 i_2 1 i_3 0 i_4 1} |j_1 0, j_2 0, \ldots, i_4 1, \ldots, j_{N_e} 0\rangle$$

$$+ V_{i_1 1 i_2 0 i_3 1 i_4 0} |j_1 0, j_2 0, \ldots, i_3 1, \ldots, j_{N_e} 0\rangle$$

$$+ V_{i_1 0 i_2 1 i_3 1 i_4 0} |j_1 0, j_2 0, \ldots, i_3 1, \ldots, j_{N_e} 0\rangle. \tag{C.12}$$

From the definition (C.8), the first term clearly gives

$$h_{L'L} \sum_{\substack{k \\ j'_k \neq i_1 \\ j'_k \neq i_2}} |j'_1 0, j'_2 0, \ldots, j'_k 1, \ldots, j'_{N_e} 0\rangle, \quad j'_k = j_k. \tag{C.13}$$

The next two terms can be shown to give,

$$h_{L'L} |j'_1 0, j'_2 0, \ldots, i_1 1, \ldots, j'_{N_e} 0\rangle \tag{C.14}$$

and the remaining terms become

$$h_{L'L} |j'_1 0, j'_2 0, \ldots, i_2 1, \ldots, j'_{N_e} 0\rangle, \tag{C.15}$$

if we use the explicit dependence of the coefficients $A_{i_1 K_1 i_2 K_2 i_3 K_3 i_4 K_4}$ on the Landau level:

$$A_{i_1 K_1 i_2 K_2 i_3 K_3 i_4 K_4} = b(i_1, i_2, i_3, i_4; q) B_{K_1 K_4}(q) B_{K_2 K_3}(-q). \tag{C.16}$$

The factors $B_{KK'}$ are given by the table:

K	K'	$B_{KK'}$
0	0	1
0	1	$-\frac{1}{\sqrt{2}}(iq_x + q_y)\ell_0$
1	0	$-\frac{1}{\sqrt{2}}(iq_x - q_y)\ell_0$
1	1	$1 - \frac{1}{2}q^2\ell_0^2$

Our assertion (C.11) is then proved by summing the terms (C.13-15).

When we now sum both sides of (C.11) over the indices i_j we get

$$\mathcal{V}^1 |L\rangle_e = \sum_{L'} h_{L'L} |L'\rangle_e \tag{C.17}$$

which when multiplied by c_L and summed over L, leads to,

$$\mathcal{V}^1|\Psi\rangle_e = \varepsilon|\Psi\rangle_e. \tag{C.18}$$

The remaining task is to show that the state $|\Psi\rangle_e$ is an eigenstate of the relative kinetic energy operator $\mathcal{K}^R = \frac{1}{2mN_e}\sum_{l<k}(\Pi_l - \Pi_k)^2$. This can be accomplished if we first note that the application of the scalar product of momentum operators yields

$$\begin{aligned}\Pi_1 \cdot \Pi_2 &|j_1 K_1 j_2 K_2\rangle \\ = \frac{\hbar^2}{\ell_0^2} C_{K_1} C_{K_2} &\left[\frac{K_2}{C_{K_1+1}C_{K_2-1}}|j_1 K_1 + 1 j_2 K_2 - 1\rangle \right. \\ &\left. + \frac{K_1}{C_{K_1-1}C_{K_2+1}}|j_1 K_1 - 1 j_2 K_2 + 1\rangle\right]\end{aligned} \tag{C.19}$$

using the explicit expressions for the single particle states. Thus the cross terms of the operator $\mathcal{K}^R$ can easily be seen to satisfy

$$\frac{1}{mN_e}\sum_{k<l}\Pi_k \cdot \Pi_l |L\rangle_e = \hbar\omega_c\left(1 - \frac{1}{N_e}\right)|L\rangle_e. \tag{C.20}$$

This state is of course also an eigenstate of the direct terms of $\mathcal{K}^R$.

Combining this last result with (C.18) completes our proof of the repeated spectrum in the finite-electron system. It should be emphasized, however, as is obvious from the treatment above, that the existence of repetition does *not* depend on any particular number of electrons and a further analysis of the terms $B_{KK'}$ reveals that it does not depend on the form of the interaction either.

Let us now turn our attention to a many-particle system following closely the SMA scheme. Let us suppose that the wave vector $|\Psi_0\rangle$ describes the true ground state of a partially filled lowest Landau level. As before we study excitations where exactly one particle is elevated to the next level. There are several possibilities to raise a particle from the ground state. For example we can excite the particles *coherently* if we give some amount of linear momentum to a particle and at the same time promote it to the next Landau level, or we can excite them *incoherently* if we first give momentum to the particles and then elevate a particle to the higher level. The first alternative leads us to the SMA excitation which is described by the wave function

$$|\Psi_C\rangle = C_C \tilde{\rho}_k^{10}|\Psi_0\rangle \tag{C.21}$$

and the second one by the wave function

$$|\Psi_I\rangle = C_I L_+ \bar{\rho}_k |\Psi_0\rangle. \tag{C.22}$$

Here $L_+ = \sum_i a_i^\dagger$ is the Landau level raising operator. The coherent excitation operator is given by

$$\bar{\rho}_k^{10} = \sum_i a_i^\dagger B_i(k). \tag{C.23}$$

The corresponding momentum creation operator on the lowest Landau level is defined by

$$\bar{\rho}_k = \sum_i B_i(k) = \sum_i \exp\left\{-ipb_i\right\} \exp\left\{-ip^* b_i^\dagger\right\}. \tag{C.24}$$

The relevant commutation relations for our purposes are

$$[a_i, a_i^\dagger] = 1, \quad [a_i, B_i] = [a_i^\dagger, B_i] = 0. \tag{C.25}$$

A straightforward application of these relations provides us with the normalization coefficients

$$\begin{aligned} C_C &= \frac{1}{\sqrt{N_e}} e^{+\frac{1}{4}|k|^2} \\ C_I &= \frac{1}{N_e} \bar{s}^{-\frac{1}{2}} \end{aligned} \tag{C.26}$$

where $\bar{s}$ is the projected static structure function (4.49,50). Similar calculation also reveals that $|\Psi_C\rangle$ and $|\Psi_I\rangle$ are indeed independent excitation modes since their scalar product

$$\langle \Psi_C | \Psi_I \rangle = \sqrt{\frac{\bar{s}}{N_e}} e^{+\frac{1}{4}|k|^2} \tag{C.27}$$

vanishes as the number of particles tends to infinity.

Our next step is to evaluate the energies. From now on we will concentrate solely on the incoherent states (C.22) since the other mode will yield the well-known SMA magnetoplasmon excitation energies (see Sect. 5.2). The state (C.22) is obviously an eigenstate of the kinetic energy operator which, with the help of the level raising operators, can be written in the form $\mathcal{K} = \hbar\omega_c \sum_i \left(a_i^\dagger a_i + \frac{1}{2}\right)$. The relevant part of the potential energy is

given by
$$\mathcal{V} = \int \frac{d^2q}{(2\pi)^2} V(q) \rho_{-q} \rho_q, \qquad (C.28)$$

where ρ is the full density operator defined by

$$\rho_q = \sum_i A_i(q) B_i(q) \qquad (C.29)$$

where

$$A_i(q) = \exp\left\{-\frac{i}{\sqrt{2}} q a_i^\dagger\right\} \exp\left\{-\frac{i}{\sqrt{2}} q^* a_i\right\}. \qquad (C.30)$$

The expectation value of the Hamiltonian in the incoherent state can now be written in the form

$$\Delta \epsilon \equiv \langle \Psi_I | \mathcal{H} | \Psi_I \rangle - E_0 - \hbar \omega_c N_e = C_I^2 \langle \Psi_0 | L_- \bar{\rho}_{-k} [\mathcal{V}, L_+ \bar{\rho}_k] | \Psi_0 \rangle. \qquad (C.31)$$

Here E_0 denotes the ground state energy. Let us now apply the commutation relation,

$$[\rho_q, L_+] = -i \frac{q^*}{\sqrt{2}} \rho_q \qquad (C.32)$$

where we have used the identity

$$[A_i(q), a_j^\dagger] = -i \frac{q^*}{\sqrt{2}} \delta_{ij} A_i(q) \qquad (C.33)$$

obtained easily from the commutation rules (C.25) and the definition (C.30). It is now a simple matter to show that the potential energy *commutes* with the Landau level raising operator, i.e.

$$[\mathcal{V}, L_+] = 0. \qquad (C.34)$$

Substituting this result and the normalization coefficient from (C.26) into (C.31) and introducing the strength factor defined earlier (4.45) as

$$\bar{f}(k) = \frac{1}{N_e} \langle \Psi_0 | \bar{\rho}_{-k} [\mathcal{V}, \bar{\rho}_k] | \Psi_0 \rangle,$$

we have

$$\Delta \varepsilon = \frac{\bar{f}(k)}{\bar{s}(k)}. \qquad (C.35)$$

This is clearly the intra-Landau level magnetoroton energy in the SMA scheme [compare (4.51)] and thus proves our assertion.

Again it should be emphasized that the validity of the formula (C.35) does not depend on the form of the interaction as long as it can be written according to (C.28), nor does it depend on any particular filling fraction.

The results we have obtained above are summarized in a simple statement: The interaction does not depend on the Landau level. This statement is however not so trivial as it at first sight appears to be. Due to projections applied to various operators it is by no means obvious that the commutation relation (C.34) holds although it is physically reasonable. It is even less obvious in the case of a finite electron system.

References

Chapter 1

1.1 K. von Klitzing, G. Dorda, M. Pepper: Phys. Rev. Lett. **45**, 494 (1980)

1.2 K. von Klitzing: Surf. Sci. **113**, 1 (1982)

1.3 K. von Klitzing: Rev. Mod. Phys. **58**, 519 (1986)

1.4 D. C. Tsui, H. L. Störmer, A. C. Gossard: Phys. Rev. Lett. **48**, 1559 (1982)

1.5 H. L. Störmer, A. M. Chang, D. C. Tsui, J. C. M. Hwang, A. C. Gossard, W. Wiegmann: Phys. Rev. Lett. **50**, 1953 (1983)

1.6 H. L. Störmer: *Advances in Solid State Physics*, ed. by P. Grosse, vol. 24 (Vieweg, Braunschweig 1984), p. 25

1.7 T. Ando, A. B. Fowler, F. Stern: Rev. Mod. Phys. **54**, 437 (1982)

1.8 H. Aoki: Rep. Prog. Phys. **50**, 655 (1987)

1.9 B. I. Halperin: Helv. Phys. Acta **56**, 75 (1983)

1.10 R. E. Prange, S. M. Girvin (eds.): *The Quantum Hall Effect* (Springer, New York, Berlin, Heidelberg 1987)

1.11 K. von Klitzing: private communication

1.12 A. M. Chang, P. Berglund, D. C. Tsui, H. L. Störmer, J. C. M. Hwang: Phys. Rev. lett. **53**, 997 (1984)

1.13 E. E. Mendez, L. L. Chang, M. Heiblum, L. Esaki, M. Naughton, K. Martin, J. Brooks: Phys. Rev. **B30**, 7310 (1984)

1.14 G. Ebert, K. von Klitzing, J. C. Maan, G. Remenyi, C. Probst, G. Weimann, W. Schlapp: J. Phys. **C17**, L775 (1984)

1.15 G. S. Boebinger, A. M. Chang, H. L. Störmer, D. C. Tsui: Phys. Rev. **B32**, 4268 (1985)

1.16 R. G. Clark, R. J. Nicholas, A. Usher, C. T. Foxon, J. J. Harris: Surf. Sci. **170**, 141 (1986)

1.17 G. S. Boebinger: In *The Physics of the Two-Dimensional Electron Gas*, ed. by J. T. Devreese, F. M. Peeters (Plenum, New York, London 1987) p. 51

1.18 R. Willet, J. P. Eisenstein, H. L. Störmer, D. C. Tsui, A. C. Gossard, J. H. English: Phys. Rev. Lett. **59**, 1776 (1987)

1.19 H. Fukuyama, P. M. Platzman, P. W. Anderson: Phys. Rev. **B19**, 5211 (1979)

1.20 D. Yoshioka, P. A. Lee: Phys. Rev. **B27**, 4986 (1983)

1.21 D. Yoshioka, B. I. Halperin, P. A. Lee: Phys. Rev. Lett. **50**, 1219 (1983)

1.22 R. B. Laughlin: In *The Quantum Hall Effect*, ed. by R. E. Prange, S. M. Girvin (Springer, New York, Berlin, Heidelberg 1987) p. 233

1.23 S. T. Chui: Phys. Rev. **B32**, 1436 (1985)

1.24 S. T. Chui, T. M. Hakim, K. B. Ma: Phys. Rev. **B33**, 7110 (1986)

1.25 F. Claro: Solid State Commun. **53**, 27 (1985)

1.26 R. B. Laughlin: Phys. Rev. Lett. **50**, 1395 (1983)

1.27 R. Tao, D. J. Thouless: Phys. Rev. **B28**, 1142 (1983)

1.28 D. J. Thouless: Phys. Rev. **B31**, 8305 (1985)

1.29 S. Kivelson, C. Kallin, D. Arovas, J. R. Schrieffer: Phys. Rev. Lett. **56**, 873 (1986)

1.30 D. H. Lee, G. Baskaran, S. Kivelson: Phys. Rev. Lett. **59**, 2467 (1987)

1.31 D. J. Thouless, Qin Li: Phys. Rev. **B36**, 4581 (1987)

1.32 S. Kivelson, C. Kallin, D. Arovas, J. R. Schrieffer: Phys. Rev. **B37**, 9085 (1988)

1.33 F. D. M. Haldane: In *The Quantum Hall Effect*, ed. by R. E. Prange, S. M. Girvin (Springer, New York, Berlin, Heidelberg 1987) p.303

Chapter 2

2.1 D. Yoshioka, B. I. Halperin, P. A. Lee: Phys. Rev. Lett. **50**, 1219 (1983)

2.2 R. B. Laughlin: Phys. Rev. Lett. **50**, 1395 (1983)

2.3 R. B. Laughlin: Surf. Sci. **142**, 163 (1984)

2.4 R. B. Laughlin: In *The Quantum Hall Effect*, ed. by R. E. Prange, S. M. Girvin (Springer, New York, Berlin, Heidelberg 1987), p. 233

2.5   D. Levesque, J. J. Weis, A. H. MacDonald: Phys. Rev. B30, 1056 (1984)
2.6   R. Morf, B. I. Halperin: Phys. Rev. B33, 2221 (1986)
2.7   R. Morf, B. I. Halperin: Z. Phys. B68, 391 (1987)
2.8   F. D. M. Haldane: Phys. Rev. Lett. 51, 605 (1983)
2.9   F. D. M. Haldane, E. H. Rezayi: Phys. Rev. Lett. 54, 237 (1985)
2.10 D. Yoshioka, P. A. Lee: Phys. Rev. B27, 4986 (1983)
2.11 D. Yoshioka, B. I. Halperin, P. A. Lee: Surf. Sci. 142, 155 (1984)
2.12 D. Yoshioka: Phys. Rev. B29, 6833 (1984)
2.13 B. I. Halperin: Helv. Phys. Acta 56, 75 (1983)
2.14 R. B. Laughlin: Phys. Rev. B27, 3383 (1983)
2.15 Yu. A. Bychkov, S. V. Iordanskii, G. M. Eliashberg: JETP Lett. 33, 143 (1981)
2.16 J. M. Caillol, D. Levesque, J. J. Weis, J. P. Hansen: J. Stat. Phys. 28, 325 (1982)
2.17 J. P. Hansen, D. Levesque: J. Phys. C14, L603 (1981)
2.18 J. G. Zabolitzky: Adv. Nucl. Phys. 12, 1 (1981)
2.19 B. Janovici: Phys. Rev. Lett. 46, 386 (1981)
2.20 J. F. Springer, M. A. Pokrant, F. A. Stevens: J. Chem. Phys. 58, 4863 (1973)
2.21 F. D. M. Haldane, E. H. Rezayi: Phys. Rev. B31, 2529 (1985)
2.22 S. M. Girvin: Phys. Rev. B29, 6012 (1984)
2.23 S. A. Trugman, S. Kivelson: Phys. Rev. B31, 5280 (1985)
2.24 V. L. Pokrovskii, A. L. Talapov: J. Phys. C18, L691 (1985)
2.25 A. Peres: Phys. Rev. 167, 1449 (1968)
2.26 G. Fano, F. Ortolani, E. Colombo: Phys. Rev. B34, 2670 (1986)
2.27 R. Morf, N. d'Ambrumenil, B. I. Halperin: Phys. Rev. B34, 3037 (1986)
2.28 F. D. M. Haldane: In *The Quantum Hall Effect*, ed. by R. E. Prange, S. M. Girvin (Springer, New York, Berlin, Heidelberg 1987), p.303
2.29 W. Duncan, E. E. Schneider: Phys. Lett. 1, 23 (1963)
2.30 Tapash Chakraborty, F. C. Zhang: Phys. Rev. B29, 7032 (1984)
2.31 F. C. Zhang, Tapash Chakraborty: Phys. Rev. B30, 7320 (1984)

2.32   H. L. Störmer, A. M. Chang, D. C. Tsui, J. C. M. Hwang, A. C. Gossard, W. Wiegmann: Phys. Rev. Lett. **50**, 1953 (1983)

2.33   M. Rasolt, F. Perrot, A. H. MacDonald: Phys. Rev. Lett. **55**, 433 (1985)

2.34   M. Rasolt, A. H. MacDonald: Phys. Rev. **B34**, 5530 (1986)

2.35   M. Rasolt, B. I. Halperin, D. Vanderbilt: Phys. Rev. Lett. **57**, 126 (1986)

2.37   F. F. Fang, W. E. Howard: Phys. Rev. Lett. **16**, 797 (1966)

2.38   A. H. MacDonald, G. C. Aers: Phys. Rev. **B29**, 5976 (1984)

2.39   F. C. Zhang, S. Das Sarma: Phys. Rev. **B33**, 2903 (1986)

2.40   Tapash Chakraborty: Phys. Rev. **B34**, 2926 (1986)

2.41   Tapash Chakraborty, P. Pietiläinen, F. C. Zhang: Phys. Rev. Lett. **57**, 130 (1986)

2.42   K. Maki, X. Zotos: Phys. Rev. **B28**, 4349 (1983)

2.43   P. K. Lam, S. M. Girvin: Phys. Rev. **B30**, 473 (1984)

2.44   E. Mendez, M. Heiblum, L. L. Chang, L. Esaki: Phys. Rev. **B28**, 4886 (1983)

2.45   A. M. Chang, P. Berglund, D. C. Tsui, H. L. Störmer, J. C. M. Hwang: Phys. Rev. Lett. **53**, 997 (1984)

Chapter 3

3.1   R. B. Laughlin: Phys. Rev. Lett. **50**, 1395 (1983)

3.2   R. B. Laughlin: Surf. Sci. **142**, 163 (1984)

3.3   B. I. Halperin: Helv. Phys. Acta **56**, 75 (1983)

3.4   R. Morf, B. I. Halperin: Phys. Rev. **B33**, 2221 (1986)

3.5   H. Fertig, B. I. Halperin: Phys. Rev. **B36**, 6302 (1987)

3.6   B. I. Halperin: Surf. Sci. **170**, 115 (1986)

3.7   Tapash Chakraborty: Phys. Rev. **B31**, 4026 (1985)

3.8   Tapash Chakraborty: Phys. Rev. **B34**, 2926 (1986)

3.9   J. M. Caillol, D. Levesque, J. J. Weis, J. P. Hansen: J. Stat. Phys. **28**, 325 (1982)

3.10   Yu. A. Bychkov, E. I. Rashba: JETP **63**, 200 (1986)

3.11   S. Kawaji, J. Wakabayashi, J. Yoshino, H. Sakaki: J. Phys. Soc. Jpn. **53**, 1915 (1984)

3.12 D. Yoshioka, B. I. Halperin, P. A. Lee: Phys. Rev. Lett. **50**, 1219 (1983)
3.13 D. Yoshioka: J. Phys. Soc. Jpn. **53**, 3740 (1984)
3.14 Tapash Chakraborty, P. Pietiläinen, F. C. Zhang: Phys. Rev. Lett. **57**, 130 (1986)
3.15 Tapash Chakraborty, P. Pietiläinen: Phys. Scr. **T14**, 58 (1986)
3.16 F. C. Zhang, S. Das Sarma: Phys. Rev. **B33**, 2903 (1986)
3.17 F. D. M. Haldane, E. H. Rezayi: Phys. Rev. Lett. **54**, 237 (1985)
3.18 G. Fano, F. Ortolani, E. Colombo: Phys. Rev. **B34**, 2670 (1986)
3.19 R. Morf, B. I. Halperin: Z. Phys. **B68**, 391 (1987)
3.20 A. H. MacDonald, S. M. Girvin: Phys. Rev. **B33**, 4414 (1986)
3.21 A. H. MacDonald, S. M. Girvin: Phys. Rev. **B34**, 5639 (1986)
3.22 B. Tausendfreund, K. von Klitzing: Surf. Sci. **142**, 220 (1984)
3.23 A. M. Chang, M. A. Paalanen, D. C. Tsui, H. L. Störmer, J. C. Hwang: Phys. Rev. **B28**, 6133 (1983)
3.24 I. V. Kukushkin, V. B. Timofeev, P. A. Cheremnykh: JETP Lett. **41**, 321 (1985)
3.25 I. V. Kukushkin, V. B. Timofeev: JETP **62**, 976 (1985)
3.26 G. S. Boebinger, A. M. Chang, H. L. Störmer, D. C. Tsui: Phys. Rev. Lett. **55**, 1606 (1985)
3.27 G. S. Boebinger, A. M. Chang, H. L. Störmer, D. C. Tsui, J. C. M. Hwang, A. Cho, C. Tu, G. Weimann: Surf. Sci. **170**, 129 (1986)
3.28 J. Wakabayashi, S. Kawaji, J. Yoshino, H. Sakaki: J. Phys. Soc. Jpn. **55**, 1319 (1986)
3.29 G. E. Ebert, K. von Klitzing, J. C. Maan, G. Remenyi, C. Probst, G. Weimann, W. Schlapp: J. Phys. **C17**, L775 (1984)
3.30 V. M. Pudalov, S. G. Semenchinskii: JETP Lett. **39**, 170 (1984)
3.31 G. S. Boebinger, H. L. Störmer, D. C. Tsui, A. M. Chang, J. C. M. Hwang, A. Y. Cho, C. W. Tu, G. Weimann: Phys. Rev. **B36**, 7919 (1987)
3.32 A. H. MacDonald, K. L. Liu, S. M. Girvin, P. M. Platzman: Phys. Rev. **B33**, 4014 (1986)
3.33 A. Gold: Europhys. Lett. **1**, 241, 479(E), (1986)
3.34 R. B. Laughlin, M. L. Cohen, J. M. Kosterlitz, H. Levine, S. B. Libby, A. M. M. Pruisken: Phys. Rev. **B32**, 1311 (1985)
3.35 R. B. Laughlin: Surf. Sci. **170**, 167 (1986)

3.36   I. V. Kukushkin, V. B. Timofeev: Surf. Sci. **196**, 196 (1988)

3.37   E. Mendez: Surf. Sci. **170**, 561 (1986)

3.38   Y. Guldner, M. Voos, J. P. Vieren, J. P. Hirtz, M. Heiblum: Phys. Rev. B**36**, 1266 (1987)

3.39   R. J. Haug, K. von Klitzing, R. J. Nicholas, J. C. Maan, G. Weimann: Phys. Rev. B **36**, 4528 (1987); R. J. Haug: Dissertation, Max-Planck-Institut, Stuttgart (1988)

3.40   R. B. Laughlin: In *Proceedings of the 17 th Int. Conf. on the Physics of Semiconductors*, ed. by D. J. Chadi, W. A. Harrison (Springer, New York, Berlin, Heidelberg 1985) p. 255

3.41   F. D. M. Haldane: Phys. Rev. Lett. **51**, 605 (1983)

3.42   B. I. Halperin: Phys. Rev. Lett. **52**, 1583, 2390(E) (1984)

3.43   R. B. Laughlin: In *The Quantum Hall Effect*, ed. by R. E. Prange, S. M. Girvin (Springer, New York, Berlin, Heidelberg 1987) p. 233

3.44   F. D. M. Haldane: In *The Quantum Hall Effect*, ed. by R. E. Prange, S. M. Girvin (Springer, New York, Berlin, Heidelberg 1987) p. 303

3.45   F. C. Zhang: Phys. Rev. B**34**, 5598 (1986)

3.46   D. Arovas, J. R. Schrieffer, F. Wilczek, Phys. Rev. Lett. **53**, 722 (1984)

3.47   A. H. MacDonald, G. C. Aers, M. W. C. Dharma-wardana: Phys. Rev. B**31**, 5529 (1985)

3.48   D. Yoshioka, P. A. Lee: Phys. Rev. B**27**, 4986 (1983)

3.49   A. H. MacDonald: Phys. Rev. B**30**, 4392 (1984)

3.50   A. H. MacDonald, D. B. Murray: Phys. Rev. B**32**, 2707 (1985)

3.51   F. C. Zhang, Tapash Chakraborty: Phys. Rev. B**34**, 7076 (1986)

3.52   R. Morf, N. d'Ambrumenil, B. I. Halperin: Phys. Rev. B**34**, 3037 (1986)

3.53   H. L. Störmer, A. M. Chang, D. C. Tsui, J. C. M. Hwang, A. C. Gossard, W. Wiegmann: Phys. Rev. Lett. **50**, 1953 (1983)

Chapter 4

4.1   F. D. M. Haldane and E. H. Rezayi: Phys. Rev. Lett. **54**, 237 (1985)

4.2   F. D. M. Haldane: Phys. Rev. Lett. **55**, 2095 (1985)

4.3   R. B. Laughlin: Physica **126B**, 254 (1984)

4.4   G. Fano, F. Ortolani, E. Colombo: Phys. Rev. **B34**, 2670 (1986)
4.5   W. P. Su: Phys. Rev. **B30**, 1069 (1984)
4.6   W. P. Su: Phys. Rev. **B32**, 2617 (1985)
4.7   D. Yoshioka: Phys. Rev. **B29**, 6833 (1984)
4.8   Q. Niu, D. J. Thouless, Y. S. Wu: Phys. Rev. **B31**, 3372 (1985)
4.9   J. Avron, R. Seiler: Phys. Rev. Lett. **54**, 259 (1985)
4.10 E. Brown: Solid State Phys. **22**, 313 (1968)
4.11 P. Maksym: J. Phys. **C18**, L433 (1985)
4.12 R. Tao, F. D. M. Haldane: Phys. Rev. **B33**, 3844 (1986)
4.13 D. Yoshioka: J. Phys. Soc. Jpn. **55**, 885 (1986)
4.14 C. Kallin, B. I. Halperin: Phys. Rev. **B30**, 5655 (1984)
4.15 D. Yoshioka: J. Phys. Soc. Jpn. **53**, 3740 (1984)
4.16 D. Yoshioka: J. Phys. Soc. Jpn. **55**, 3960 (1986)
4.17 E. H. Rezayi: Phys. Rev. **B36**, 5454 (1987)
4.18 M. Rasolt, A. H. MacDonald: Phys. Rev. **B34**, 5530 (1986)
4.19 S. M. Girvin, A. H. MacDonald, P. M. Platzman: Phys. Rev. Lett. **54**, 581 (1985)
4.20 S. M. Girvin, A. H. MacDonald, P. M. Platzman: Phys. Rev. **B33**, 2481 (1986)
4.21 R. P. Feynman: *Statistical Physics* (Benjamin, Reading Mass. 1972) Chap. 11
4.22 R. P. Feynman, M. Cohen: Phys. Rev. **102**, 1189 (1956)
4.23 S. M. Girvin, T. Jach: Phys. Rev. **B29**, 5617 (1984)
4.24 A. Kallio, J. Kinaret, M. Puoskari: to be published
4.25 M. Saarela: Phys. Rev. **B35**, 854 (1987)
4.26 H. C. A. Oji, A. H. MacDonald, S. M. Girvin: Phys. Rev. Lett. **58**, 824 (1987)
4.27 W. P. Su and Y. K. Wu: Phys. Rev. **B36**, 7565 (1987)
4.28 D. Yoshioka: J. Phys. Soc. Jpn. **56**, 1301 (1987)
4.29 E. Gornik, R. Lassing, G. Strasser, H. L. Störmer, A. C. Gossard: Surf. Sci. **170**, 277 (1986)

Chapter 5

5.1   C. Kallin, B. I. Halperin: Phys. Rev. B30, 5655 (1984)

5.2   C. Kallin, B. I. Halperin: Phys. Rev. B31, 3635 (1985)

5.3   B. A. Wilson, S. J. Allen, D. C. Tsui: Phys. Rev. Lett. 44, 479 (1981)

5.4   B. A. Wilson, S. J. Allen, D. C. Tsui: Phys. Rev. B24, 5887 (1981)

5.5   G. L. J. A. Rikken, H. W. Myron, P. Wyder, G. Weimann, W. Schlapp, R. E. Horstman, J. Wolter: J. Phys. C18, L175 (1985)

5.6   Z. Schlesinger, S. J. Allen, J. C. M. Hwang, P. M. Platzman, N. Tzoar: Phys. Rev. B30, 435 (1984)

5.7   Z. Schlesinger, W. I. Wang, A. H. MacDonald: Phys. Rev. Lett. 58, 73 (1987)

5.8   W. Kohn: Phys. Rev. 123, 1242 (1961)

5.9   E. Batke, D. Heitman, J. P. Kotthaus, K. Ploog: Phys. Rev. Lett. 54, 2367 (1985)

5.10 A. H. MacDonald: J. Phys. C18, 1003 (1985)

5.11 Yu. A. Bychkov, S. V. Iordanskii, G. M. Eliashberg: JETP Lett. 33, 143 (1981)

5.12 A. H. MacDonald, H. C. A. Oji, S. M. Girvin: Phys. Rev. Lett. 55, 2208 (1985)

5.13 H. C. A. Oji, A. H. MacDonald: Phys. Rev. B33, 3810 (1986)

5.14 P. Pietiläinen, Tapash Chakraborty: Europhys. Lett. 5, 157 (1988)

5.15 P. Pietiläinen: Dissertation, University of Oulu (1988); Phys. Rev. B (15 August 1988).

Chapter 6

6.1   V. L. Pokrovskii, A. L. Talapov: JETP Lett. 42, 80 (1985)

6.2   V. L. Pokrovskii, A. L. Talapov: JETP 63, 455 (1986)

6.3   F. C. Zhang, V. Z. Vulovic, Y. Guo, S. Das Sarma: Phys. Rev. B32, 6920 (1985)

6.4   E. H. Rezayi, F. D. M. Haldane : Phys. Rev. B32, 6924 (1985)

6.5   S. M. Girvin, A. H. MacDonald, P. M. Platzman: Phys. Rev. B33, 2481 (1986)

6.6   R. Tao, F. D. M. Haldane: Phys. Rev. B33, 3844 (1986)

6.7 F. C. Zhang: Phys. Rev. B**34**, 5598 (1986)
6.8 A. H. MacDonald: Phys. Rev. B**30**, 3550 (1984)
6.9 A. H. MacDonald, S. M. Girvin: Phys. Rev. B**33**, 4009 (1986)
6.10 F. D. M. Haldane: In *The Quantum Hall Effect*, ed. by R. E. Prange, S. M. Girvin (Springer, New York, Berlin, Heidelberg 1987) p. 303
6.11 N. d'Ambrumenil, A. M. Reynolds: J. Phys. C**21**, 119 (1988)
6.12 R. G. Clark, R. J. Nicholas, A. Usher, C. T. Foxon, J. J. Harris: Surf. Sci. **170**, 141 (1986)
6.13 A. H. MacDonald, S. M. Girvin: Phys. Rev. B**34**, 5639 (1986)
6.14 V. Kalmeyer, R. B. Laughlin: Phys. Rev. Lett. **59**, 2095 (1987)
6.15 B. I. Halperin: Helv. Phys. Acta **56**, 75 (1983)
6.16 D. Yoshioka: Phys. Rev. B**29**, 6833 (1984)
6.17 F. D. M. Haldane: Phys. Rev. Lett. **55**, 2095 (1985)
6.18 Y. Kuramoto, R. R. Gerhardts: J. Phys. Soc. Jpn. **51**, 3810 (1982)
6.19 Tapash Chakraborty and P. Pietiläinen: to be published
6.20 G. Fano, F. Ortolani, E. Tosatti: Nuovo Cimento **9D**, 1337 (1987)
6.21 G. E. Ebert, K. von Klitzing, J. C. Maan, G. Remenyi, C. Probst, G. Weimann, W. Schlapp: J. Phys. C**17**, L775 (1984)
6.22 R. G. Clark, R. J. Nicholas, J. R. Mallett, A. M. Suckling, A. Usher, J. J. Harris, C. T. Foxon: In *Proc. of the Eighteenth Int. Conf. on the Phys. of Semicond.*, ed. by O. Engstrom (World Scientific, Singapore 1987) p. 393
6.23 R. Willet, J. P. Eisenstein, H. L. Störmer, D. C. Tsui, A. C. Gossard, J. H. English: Phys. Rev. Lett. **59**, 1776 (1987)
6.24 F. D. M. Haldane, E. H. Rezayi: Phys. Rev. Lett. **60**, 956 (1988)
6.25 Tapash Chakraborty, P. Pietiläinen: Phys. Rev. Lett. **59**, 2784 (1987)
6.26 W. L. Bloss, E. M. Brody: Solid State Commun. **43**, 523 (1982)
6.27 S. Das Sarma, J. J. Quinn: Phys. Rev. B**25**, 7603 (1982)
6.28 Tapash Chakraborty, C. E. Campbell: Phys. Rev. B**29**, 6640 (1984)
6.29 P. B. Visscher, L. M. Falicov: Phys. Rev. B**3**, 2541 (1971)
6.30 G. Fasol, N. Mestres, H. P. Hughes, A. Fischer, K. Ploog: Phys. Rev. Lett. **56**, 2517 (1986)
6.31 A. Pinczuck, M. G. Lamont, A. C. Gossard: Phys. Rev. Lett. **56**, 2092 (1986)

6.32   P. Maksym: J. Phys. C18, L433 (1985)

6.33   H. C. A. Oji, A. H. MacDonald, S. M. Girvin: Phys. Rev. Lett. 58, 824 (1987)

Appendix A

A.1   R. B. Dingle: Proc. Roy. Soc. A211, 500 (1952)

A.2   L. D. Landau, E. M. Lifshitz: *Quantum Mechanics*, (Pergamon, Oxford 1977) p. 458

A.3   S. M. Girvin, T. Jach: Phys. Rev. B28, 4506 (1983)

A.4   S. M. Girvin, T. Jach: Phys. Rev. B29, 5617 (1984)

A.5   V. Bergmann: Rev. Mod. Phys. 34, 829 (1962)

Appendix B

B.1   J. M. J. van Leeuwen, J. Groeneveld, J. de Boer: Physica 25, 792 (1959)

B.2   T. Morita, K. Hiroike: Prog. Theor. Phys. 23, 1003 (1960)

B.3   E. E. Salpeter: Ann. Phys. 5, 183 (1958)

B.4   J. M. Caillol, D. Levesque, J. J. Weis, J. P. Hansen: J. Stat. Phys. 28, 325 (1982)

B.5   K. Hiroike: Prog. Theor. Phys. 24, 317 (1960)

B.6   Tapash Chakraborty: Phys. Rev. B26, 6131 (1982)

B.7   Tapash Chakraborty, P. Pietiläinen: Phys. Rev. Lett. 49, 1034 (1982)

Subject Index

Activation energy 65–67, 71, 73
Activation gap 65
    as a function of mobility 69
    measurements of the 65–67
Angular momentum
    azimuthal 54
    conservation of 19, 100
    eigenstate of 18, 100, 144
    kinetic 25
Articulation point 148
Aspect ratio 13, 90

Backflow 102–103
Basis states 14, 16, 75
    construction of 92
Boson ladder operators 146

Center-of-mass translation 88
Charge-density-wave 7, 106–107
Chemical potential 7, 49–50, 71
Classical plasma
    analogy 19, 53, 60
    charge neutrality condition of 19
    crystallization transition 37
    dimensionless parameter of 20
    Hamiltonian 19
    inhomogeneous 48
    ion-disk radius of 20
    one-component 20–21, 41
    two-component 32, 41, 46, 155
Collective excitation 83
    energy 84–85, 93–96, 101, 106–107, 133, 135, 136, 139
    in higher-Landau levels 129–131
Conductivity 1, 4
    thermal activation of 65
Continuity condition 102–103
Correlation function
    pair 20–21, 29, 33, 41–43
    three-body 48

Crystallization transition 37
Cusp in ground state energy 7, 15, 49, 79, 124, 132
Cyclotron energy 6, 65
Cyclotron frequency 11
Cyclotron resonance 109, 117

Debye length 47
Definition of distance
    chord 27
    great circle 27
Degeneracy of the ground state 85
Density operator 97
    projected 99
Density-wave spectrum 84–85, 93–96, 101
Density-wave excited state 97–98
Diagrams
    composite 149
    elementary 150
    irreducible 147
    linked 148
    (non) nodal 150
    simple 149
Disorder
    effect on the activation gap 70–71
Drift velocity 7, 39

Electron-hole symmetry 16, 23
Elementary excitations 39, 77, 83, 110
Energy
    approximate formula for 22, 30
    interaction 20
Energy gap 29, 49, 55, 59, 63, 100–101
    condition for 100
    quasiparticle-quasihole 39, 49, 55, 63, 93
Even-denominator filling fractions 131, 137
    experimental observation of 135–136
Exact diagonalization of the Hamiltonian 14, 50–51

Excitations
    coherent 114, 159
    incoherent 117, 159–160
Excitation energy
    at any hierarchy level 76
Excitation spectrum
    finite-size studies of 84–85, 93–97
    in single-mode approximation 101
    in a superlattice 105

Fang-Howard variational function 35
Feynman theory 97–98
Filled Landau level
    elementary excitations in a 110
    energy of a 20
    pair-correlation function for a 20
Filling factor (fraction) 6, 14, 19, 32
    as a continued fraction 76
    observed in FQHE 3
Finite-thickness parameter 35–36
Flux quantum 6
Fractional statistics 77, 79

Hall conductivity 1
    quantization of 1–2
Harmonic oscillator 11
Hartree-Fock approximation 15, 115
Higher order filling factors 73
    trial wave function approach of 81
Hypernetted-chain method
    one-component 20–21, 147
    two-component 33, 41, 46, 155–156

Impurity interaction 122
    delta-function (short-range) 121, 125
Impurity limit 41–42
Impurity-plasma interaction 46
Impurity strength 123
Incompressibility 7, 100
Integer quantum Hall effect 1
Inter-Landau level excitations 109
    in finite-size studies 117
Intra-Landau level excitations 97
    in finite-size studies 83, 85

Laughlin wave function
    for higher Landau levels 127
    in disk geometry 18
    in spherical geometry 26
Landau level 11
    degeneracy of 12, 18–19, 25
Layered electron systems 137
    excitation spectrum of 138–139

Magnetic exciton 110
    energy 111
    wave function of 110
Magnetic length 6
    for the quasiparticle state 76
    modified 27
Magnetic translation 48
    operators 47, 86–87
Magnetoroton band in a superlattice 104–105
Magnetoroton mode 101
    repetition of 117–118, 156
Magnetoplasmon mode 109, 112, 117, 160
Mobility threshold 65, 70
Monte Carlo results for
    ground state energy 30–32
    one-particle density 56
    pair-correlation functions 29, 31
    quasihole creation energy 57
    quasiparticle creation energy 59
Multilayer structure 104, 137

One-half filled Landau level
    ground state energy of 132–133
    pair-correlation functions in 134
One-particle density 45
    integrated by parts 46–47
Oscillator strength 98
    projected 99

Particle excess 43, 57
Periodic rectangular geometry 13
    symmetry analysis 86
Pseudopotential parameters 27–28
Phantom point charge 41

Quasiexciton 84, 93, 95
Quasihole 40
    creation energy 42–43
    wave function 40

Quasiparticle 44
    creation energy 46
    spin-reversed 50, 64
    trial wave function (alternative) of 59
    wave function 44
Quasiparticle and quasihole size 47

Ray representation 87
Resistivity 4
Roton minimum 98, 101

Single-particle eigenstates
    in Landau gauge 12–13, 90
    in symmetric gauge 18, 144
Single-mode approximation 99
    for inter-Landau level excitations 112
    for intra-Landau level excitations 97
Specific heat 108
Spherical geometry
    ground state energy in 26
    ground state wave function in 24–25
    quasiparticle and quasihole in 54
Spin reversed quasiparticles
    finite-size studies of 50
    in spherical geometry 64
    wave function for 53

Spin-wave dispersion 94, 111
Spinor variables 25
Static susceptibility 125
Structure function 98–100
    dynamic 98
    projected static 99

Two-component systems 32, 106–107
Translational symmetry 85
    analysis 86–90

Valley waves 35
Vector potential
    Landau gauge 10, 86
    symmetric gauge 18, 87, 143
    in spherical geometry 25
Visscher-Falicov model 137
Vortex 40

Wigner crystal 15
    correlated 37
    radial distribution function 21–22

Zeeman energy 33
Zeros of the wave function 40